how to make biodiesel

Dan Carter

Dave Darby

Jon Hallé

Phillip Hunt

LILI

Published February 2005 by

Low-Impact Living Initiative
Redfield Community, Winslow, Bucks, MK18 3LZ, UK
+44 (0)1296 714184

lili@lowimpact.org

www.lowimpact.org

Printed in Great Britain by
Lightning Source, Milton Keynes

ISBN 0-9549171-0-3

contents

about the authors

Jon Hallé and Dan Carter

are directors of GoldenFuels Ltd, a biodiesel production cooperative in Oxfordshire, UK: www.goldenfuels.com. Jon and Dan tutor on LILI's biodiesel course.

Dave Darby

is director of LILI.

Phillip Hunt

was a co-founder of LILI, and now works for the National Energy Foundation: www.nef.org.uk.

disclaimer

It is up to you to check that your engine is biodiesel compatible, your storage of methanol and waste vegetable oil is secure, and your reactor and production techniques are safe. Anyone using the methods described in this manual does so at their own risk. LILI assumes no responsibility for damage to persons or property caused by biodiesel production using this manual as a guide.

introduction

This book contains pretty much everything you need to know to get started with biodiesel – what it is exactly, why it's good for the environment, the chemistry involved in making it, how to make small experimental batches, how to build a reactor and make larger batches, paying duty, and any legal / environmental / planning requirements you need to think about. It has was originally written to accompany LILI's 'how to make biodiesel' course. The course gives participants an opportunity to get hands-on experience, to ask any tricky questions they like, and to make like-minded contacts.

Biodiesel offers us an amazing opportunity to improve the environmental and ethical performance of our vehicles, which up to now have been one of the hardest aspects of everyday life to improve. It makes a refreshing change to be able to promote a green solution that does not involve daunting levels of investment or lifestyle change.

Biodiesel will run in any Diesel vehicle. Different vehicles require different levels of fuel quality; some will run on very simply-made fuel, others need a more refined technique. According to the US National Biodiesel Board, biodiesel has performed similarly to mineral diesel over 50 million road miles, during which it was found to have similar horsepower, torque and fuel consumption, but with better lubricating properties, extending the life of engine parts.

Biodiesel makes financial sense. You can buy biodiesel for roughly the same price as mineral diesel or make it yourself, potentially for less.

Biodiesel is perfectly legal, as long as you pay fuel duty to Customs and Excise. You may remember the recent arrests in Wales by the 'Frying Squad', as many papers termed it. The biodieselers' offence was not their use of biodiesel in itself; it was their not having paid fuel duty on the biodiesel. Incidentally, they were not really making what we call 'biodiesel' either (someone on a course labelled it 'boyodiesel'). Later in this manual we will look at the different ways of running a vehicle on plant oils.

Biodiesel can be made without the need for elaborate equipment. In this book we cover everything you need to make biodiesel on a small scale and a lot that will be useful to those considering a larger operation or wishing to use fuel made by others.

Biodiesel is recognised by the Government as an environmentally-friendly fuel. Gordon Brown knocked 20p per litre off the duty for biodiesel in the budget of 2002. The lower rate of duty is dependent on producing fuel to certain specifications that are covered in detail in this manual.

Biodiesel brings a range of benefits - for the environment, human health, vehicles, and local economies. Its greatest strength is that it is something you can do yourself. Fuel supply doesn't have to be in the hands of a tiny number of enormous companies. This power (and the wealth it brings) can be distributed to small and medium-sized companies, co-ops, farmers and individual producers.

In these times of climate change, rising fuel prices, instability in the Middle East, and the possibility of fuel protestors blockading refineries and ports, it might be a very good idea to have a biodiesel reactor in your garage.

The first part of this manual takes a broad look at the world of biodiesel. We will start by defining biodiesel. Next we will look at the reasons for running on biodiesel: what are the benefits over mineral diesel. Then we will have a closer look at the Diesel engine: how it works and how we can run it on biodiesel. A primer on the basic chemistry of biodiesel ends this section.

The second part of the book is devoted to the practicalities of making and using biodiesel. We go through the process of making a mini-batch of biodiesel to explain the basic steps. Next we look at a small-scale reactor to investigate the issues around scaling up production. Finally we look at the legal realities of making, using and selling biodiesel.

what is biodiesel?

We use the following definition of biodiesel:

Biodiesel is a fuel for conventional Diesel engines made from plant or animal oils or fats that have been chemically transformed into alkyl esters.

Lets take each part of this definition and break it down.

Biodiesel is a fuel for conventional Diesel engines...

Diesel engines are compression-ignition engines of the type most successfully developed by Rudolph Diesel. The fuel which we are used to calling 'diesel fuel' is just one of the possible fuels for this type of engine. In this manual we will call this fuel 'petrodiesel' or 'mineral diesel'. It is also called 'dinodiesel' in the biodiesel community, although it is made not from dinosaurs but from ancient plants, mostly algae. Diesel engines have been made which will run on milk powder, coal dust, or straight vegetable oil. However the most common form of Diesel engine is designed to run on petrodiesel. This is the type of engine in most cars and commercial vehicles. When we make biodiesel, we must make a fuel that is suitable for these engines. Note that it is not possible to run petrol engines on petrodiesel or biodiesel.

...made from plant or animal oils or fats...

Biodiesel can be made from a very wide range of raw materials or 'feedstocks'. They range from animal fats or tallows to vegetable oils including the common rapeseed, sunflower and peanut oils. Palm oil and coconut oil can also be used, as can tropical oil-bearing shrubs like jatropha. You can even make biodiesel from oil extracted from algae (cutting out millions of years of underground processing). To maximise the green advantage of biodiesel we make it from waste vegetable oil which has been used for deep-frying. In this manual we refer to this as WVO (waste vegetable oil). It may contain small amounts of animal fats, and many of the techniques we describe can be adapted to work with pure animal fats.

... that have been chemically transformed into alkyl esters.

It is possible to make a Diesel engine which will run directly on many plant oils. In fact one of Rudolf Diesel's first engines was exhibited at the Paris exhibition of 1902 running on pure peanut oil. Even today, Elsbett of Germany make pure vegetable oil burning engines which are often used by German taxi drivers. However, conventional Diesel engines will generally not run well on pure vegetable oil. The reason for this can be summed up in one word: viscosity.

Viscosity: The "gloopiness" or "stickiness" of a fluid; its resistance to flow

Vegetable oil is much more viscous than petrodiesel, and conventional Diesel engines simply aren't designed to deal with this. There are two ways around this problem:

1. modify the engine to accept more viscous fuel.

This is a good option in its way, but it has some major downsides. On many engines, particularly recent ones, it is impractical. On older engines it may be possible but costly, in the region of £2000. This option is not one which we cover in this book.

2. modify the fuel to reduce its viscosity

This is a more practical option for nearly all situations. Once we have decided to take this route, there are three major options:

- heating the oil: in this option we modify the fuel system of the vehicle so that the pure vegetable oil is heated on its way to the engine. This reduces its viscosity to nearer that of the petrodiesel the engine was designed for. This option is far more realistic than actually modifying the engine. However, it can still involve quite a large investment in one vehicle, around £500. It is also impractical for many more recent Diesels. This fuel is generally known as SVO (straight vegetable oil). It is not what we call biodiesel.

- mixing the oil with a thinner: there are a number of variations on this technique, which involves making a 'micro-emulsion' between vegetable oil and a solvent - for example white spirit, methanol, kerosene etc. This has the virtue of simplicity, however there is little research into the effects of this sort of blend on environmental impacts and engine life. This sort of fuel is known by a number of terms including emulsion, and MWVF (modified waste vegetable fat). In press reports it is sometimes called biodiesel. However it is not what we call biodiesel

- chemically transforming the oil to make it thinner: vegetable oil is viscous because its molecules are built around glycerol. If you chemically alter the vegetable oil to replace the glycerol with something less viscous you get a fuel which fits the specifications of conventional Diesel engines. The chemical process is known as transesterification. The glycerol is replaced by a simple alcohol such as methanol or ethanol to produce a product known as alkyl esters. This process is the subject of this manual and produces what we call biodiesel.

Remembering the key issue of viscosity will make it easier to remember the distinction between the various ways of running a Diesel engine on vegetable oils, which are often confused by beginners and journalists. In appendix 2 there is some more information about the non-biodiesel options.

Biodiesel is sometimes called RME (rape methyl ester) when it is made from new rapeseed oil (the most common feedstock in large European biodiesel factories). A more general term is FAME (fatty acid methyl esters), which covers any feedstock. Both terms assume that methanol is used as the replacement for glycerol in the reaction, though ethanol and isopropyl alcohol can also be used in theory.

different kinds of plant oils

Many types of plant oil can be used to make biodiesel, including coconut, peanut, sunflower, rapeseed, oil palm, soya bean, linseed, olive and hemp.

Rapeseed is called canola in the US. It's what's growing in the bright yellow fields you see in the summer. Rapeseed oil has no saturated fat, which is very good, as this is what makes oil solidify (sunflower oil has 8% saturated fat). So rapeseed oil is probably best for making biodiesel. Waste cooking oil from takeaways, or oil whose label just says 'vegetable oil' or 'cooking oil' is often rapeseed oil (other oils such as sunflower or olive will state which plants they are from). It has an edible version and an industrial version, sometimes called HEAR (high erucic acid rape).

Rape yields around 2 tonnes of oil per hectare (pressed from 4 tonnes of harvested seed), but also produces 4 tonnes of straw per hectare, which many boilers can now use as a fuel.

This is a good yield, but the champion is the African Palm. Palm oil can yield 5 tonnes of oil per hectare - but not in the European climate. Also, palm oil is solid at room temperature.

Rape is the clear winner in temperate regions. It grows well in a temperate climate; it has double the yield of most of the competition, and is liquid at room temperature.

The future of biodiesel production probably lies with algae though. Algae could produce (in ponds) up to ten times more oil per hectare than rapeseed. And algae love carbon dioxide; they gobble it up so fast that they could be of huge assistance in reducing global warming. They could even utilize the carbon dioxide emissions from industry. Each factory could have an algae pond to eat up its carbon emissions and produce oil. Algae can even grow in salt water or wastewater, so they don't need to compete with agriculture for fresh water.

waste vegetable oil

Throughout this manual we are focusing on making biodiesel from waste vegetable oil (WVO). This is oil that has previously been used for deep-frying and is disposed of as a waste product by catering outlets. The techniques we describe for making biodiesel apply in most cases to both WVO and fresh vegetable oils.

In the UK, using WVO is the best way to maximise the potential of biodiesel as a greener and cheaper fuel. Utilizing a waste resource saves energy and reduces pressure on landfill. Previously, waste cooking oil has gone to make cattle feed, but it meant that cattle were eating some beef residue in the oil which could transmit BSE. The EU implemented a Europe-wide ban on waste oil going to cattle feed. The UK government successfully applied for a deferment for two years, on the basis that at the time there was nowhere else for the used oil to go other than illegally disposing of it in drains or in landfill.

The deferment expired on 31st October 2004. By reducing the duty on biodiesel in the 2002 budget, it was hoped that a fledgling biodiesel industry in the UK would soak up the waste cooking oil by the time the deferment ended.

The UK produces 100,000 to 200,000 tonnes of waste oil per year. The reduction of 20p per litre for biodiesel makes biodiesel production viable with used oil. We need approximately another 10p per litre reduction to make production viable with new oil.

biofuels and fossil fuels

Both biofuels and fossil fuels are ultimately derived from living things. The distinction between biofuels and fossil fuels is essentially one of time. Fossil fuels come from plants (making coal) and tiny sea creatures (making oil and gas) that died millions of years ago, and were laid down in coal deposits, and in oil and gas fields. The key difference is that because of the time scale involved in the creation of fossil fuels, they are in effect finite – i.e. not renewable in our lifetimes (or indeed a million lifetimes). When they're gone, they're gone.

Biofuels on the other hand, are renewable. We can just plant more, and if we plant them as fast as we use them, they are a sustainable resource.

Biofuels can be in liquid form, like biodiesel, or can be burnt solid, like wood. Boilers have been developed that can take straw, woodchips or pellets, or miscanthus (a species of tall, heavy grass). In this form they are known as biomass. Plant matter can also be gasified, and biogas (methane) from landfill, or decomposing organic matter, is also a biofuel.

Bioethanol (100%) has been used successfully in vehicles, notably in Brazil, where it is made from sugar cane. It is produced in liquid form for use in petrol vehicles. In the long term, this may prove more efficient than producing biodiesel, but equipment will have to be changed – modifications are needed to petrol engines, and it can't be made very cheaply in the UK. This is not the case with biodiesel, and so production of a biofuel that can be used in existing equipment is very useful, even if it turns out to be short term. It also gets people used to the idea of biofuel, which involves a completely different way of thinking (cyclical, renewable) from fossil fuels (linear, non-renewable).

benefits of biodiesel

In this section we'll look at the environmental and other problems caused by the production and use of mineral diesel, and compare it to the production and use of biodiesel.

First we'll look at global warming, then atmospheric pollution and acid deposition (i.e. 'acid rain', although the acids can also be deposited dry) – which is the falling to the ground of acidic pollutants in the atmosphere.

Then we'll look at the other environmental benefits of using biodiesel, as well as benefits for human health and for vehicles.

CO_2 emissions and global warming

The US Department of Energy reports that production and use of biodiesel represents a reduction in CO_2 emissions of over 75% compared to mineral diesel. So although it's not a completely carbon-neutral fuel, it's close.

what is global warming?

- it's the heat from the sun trapped by gases in the atmosphere, which absorb reflected long-wave radiation (heat from the sun is both long- and short-wave; short-wave radiation passes through greenhouse gases, but once it bounces off the earth, sea or plants, it is long-wave, and is trapped in the earth's atmosphere by greenhouse gases)
- greenhouse gases include water vapour, CO_2, methane, nitrogen oxides, and some CFCs
- these gases are produced naturally by the respiration of animals, volcanic activity, forest fires and evaporation
- the amount of greenhouse gases is increasing rapidly because of human activity – especially from the burning of fossil fuels, livestock and removal of forests and biomass, which take in CO_2
- fossil fuels hold the carbon that millions of years ago used to be in the air. Plants and microscopic creatures took the carbon from the air, died, and

stored the carbon in coal, oil and gas under the ground. We're now using all the fossil fuels and putting that carbon back into the air

the 'greenhouse' gases

water vapour

- is responsible for trapping most infrared radiation bouncing off the earth, but its importance is reduced by the fact that clouds reflect a lot of solar radiation back out to space too

methane (CH_4)

- (from human activity) – 100 million tonnes from livestock; 100 million tonnes from rice paddies; 10-100 million tonnes from biomass burning
- CH_4 is destroyed in the atmosphere via reactions with the hydroxyl radical OH. ($CH_4 + OH = CH_3 + H_2O$)
- this means that the average lifetime of CH_4 in the atmosphere is 8 years

nitrogen oxides (NOx)

- building up in the atmosphere at a rate of 3 million tonnes per year
- mainly from burning fossil fuels
- destroyed in the atmosphere by solar radiation
- their average lifetime in the atmosphere: 150 years

CFCs

- CFC11 and CFC12 are the 'greenhouse' CFCs
- much smaller amounts than other greenhouse gases, so they have a relatively small effect – even though they have the biggest effect per molecule
- they also damage the ozone layer
- they come from refrigerants
- they are also destroyed by solar radiation

carbon dioxide (CO_2)

- in pre-industrial times, CO_2 in the atmosphere was 290ppm (parts per million); now it is 350ppm, and increasing at around 0.3% per year

- 7 billion tonnes of CO_2 is put into the atmosphere by human activity each year. Five billion tonnes is from burning fossil fuels, and 2 billion from deforestation
- the oceans absorb 2 billion tonnes per year, land plants absorb 2 billion, and 3 billion tonnes accumulates in the atmosphere
- it is not destroyed in the atmosphere by solar radiation or by reactions with anything else

is it really increasing?

- yes, despite the Mount Pinatubo eruption, which reduced the amount of solar radiation reaching the earth for two years in 1990s
- the average temperature at the earth's surface has increased by 0.5°C in the last 100 yrs.
- most scientists predict a rise of between 1°C and 4°C in the next 100 years (there is only a 5°C difference in average temperatures between now and last ice age)

what problems does it cause?

- the important point is the speed of global warming; warming (or cooling) over millions of years is not a problem, as plants and animals have time to adapt. When humans, in less than 200 years, take all the carbon trapped under the ground in fossil fuels and put it into the atmosphere, then temperatures change so rapidly that plants and animals don't have time to adapt – evolution doesn't work that quickly. And plants can't move north (or south in the southern hemisphere) quickly enough either, so they become extinct
- the last major extinction event happened around 250 million years ago at the end of the Permian period, when 95% of all species became extinct, and it took 150 million years to restore previous levels of biodiversity. Scientists think that the culprit was a giant volcano in Siberia (where volcanic deposits have been discovered many miles thick over millions of square miles), which raised the average global surface temperature by around 5°C in a relatively short time. This is similar to the increase predicted

by most scientists in the next century, if current trends continue
- there will be lower overall global rainfall; biomass is directly related to overall rainfall, so there will be a global biomass reduction, leading to more extinctions of species
- for humans it will mean coastal flooding, drought, forest fires, and an increase in tropical pests and diseases
- most important for humans will be desertification and famine – especially in Sub-Saharan Africa
- global warming could ironically make Britain much colder by stopping the Gulf Stream. The situation now: the Gulf Stream is warmer than the surrounding water when it reaches the North Atlantic. Warmer water contains more salt, and as it evaporates more quickly, salt concentrations increase even more. Salty water is heavier, so it sinks, and pulls more water across the Atlantic from the Caribbean. But the Greenland ice shelf is melting much faster due to global warming, and this dilutes the salt water which doesn't sink so easily. The Gulf Stream is already slowing down rapidly
- it is extinctions that could eventually cause the biggest problem for humans. The UN Environment Programme and the International Botanical Congress predict a loss of up to 60% of all species of plants and animals in the coming century. We don't know if we can survive this kind of a loss, especially because, as species are so interconnected, a 60% loss could snowball to a much higher loss. So another massive extinction event could take us with it this time

what can we do about it?

- use renewable energy (solar hot water, photovoltaics, wind generators, wood stoves, and of course, biodiesel)
- use less energy (by consuming less, making changes to our lifestyles, and by using energy efficient appliances)
- stop deforestation

emissions of pollutants

In this section we will look at the various pollutants from road traffic, especially from diesel vehicles, and the problems they cause for the environment and human health. We will then compare emissions from vehicles using mineral diesel and biodiesel.

Since the industrial revolution, air pollution has been largely from industry, and also domestic sources, such as coal fires. In the West, the decline of manufacturing industry, newer cleaner technologies, and a move from coal to natural gas in homes, has meant a reduction in pollution from these sources.

Although the growing industrial base in the developing world still contributes to atmospheric pollution, the main culprits now are power generation, and overwhelmingly, vehicles and the internal combustion engine (electric vehicles don't necessarily help – emissions just move from vehicles to power stations unless the electricity is generated from renewable sources).

We list below the environmental and health problems associated with specific pollutants, but there are further problems as those pollutants chemically react with each other and with sunlight in the atmosphere to produce secondary problems (acid rain, low-level ozone). Historically, air pollution tended to be local, in the vicinity of urban and industrial areas. Often, smog engulfed major cities, due to the burning of fossil fuels. This problem is now confined to developing countries, but now the effects of vehicle emissions can be found throughout the atmosphere, and the environmental damage is global. Secondary problems often occur far from the source of the initial pollutants.

Emissions from vehicles are becoming worse, largely due to the massive increase in vehicle numbers, both in developed and developing countries. As China shuns the bicycle for the internal combustion engine for example, we can expect hundreds of millions of extra vehicles in use, with very few enforceable emissions controls. And yet how can we in the West criticise this trend, when they have such a long way to go before they reach the level of vehicle ownership in the West, and especially North America?

nitrogen oxides (NOx)

what are they?

They are mainly the gases nitric oxide (NO) and nitrogen dioxide (NO_2); very small amounts of nitrous oxide (N_2O – or laughing gas) are emitted.

Nitrogen oxides are mainly emitted in the form of nitric oxide, which is then oxidized in the atmosphere to nitrogen dioxide, by reacting with ozone.

They are formed from all types of high-temperature combustion processes, from the oxidization of nitrogen from either the fuel itself or from the surrounding air.

sources

45% of NOx emissions in the UK are from vehicles. However, because emissions from power stations are from tall chimneys, 75% of NOx concentrations at ground level are from vehicles.

problems

Nitrogen dioxide causes damage to plants and ecosystems, and is a respiratory irritant, especially with asthmatics.

Although nitrogen oxides are damaging in themselves, they don't persist in the atmosphere – they cause more environmental problems via secondary pollutants. Nitrogen dioxide will react with hydrocarbons in sunlight to produce ozone. Also, after only one day in the atmosphere, nitrogen oxides will begin to convert to nitric acid, which is then deposited as acid rain.

sulphur dioxide (SO_2)

what is it?

It is a gas which is acid and very corrosive; it combines with water vapour in the atmosphere to produce acid rain. It is formed by the oxidation of sulphur impurities in fuels during their combustion.

sources

In the UK, over 85% of SO_2 emissions are from power stations and industry. As the number of coal fires have fallen, SO_2 concentrations in urban areas have fallen too.

Petrol engines emit virtually no SO_2, but Diesel engines do. As lower-sulphur fuel has been introduced, emissions from Diesel engines have fallen too.

problems
SO_2 can be deposited wet, as acid rain, or dry, and in both cases can damage vegetation and buildings, and degrade soils and watercourses. It is associated with asthma and chronic bronchitis.

carbon monoxide (CO)

what is it?
Carbon monoxide (CO) is a toxic gas that is emitted into the atmosphere as a result of combustion processes, and is also formed by the oxidation of hydrocarbons and other organic compounds.

sources
In developed countries' urban areas, CO is almost entirely (around 90%) from vehicle emissions. The slower the vehicles, the more the CO emissions, which explains the high concentrations in urban areas, where vehicles are slower-moving.

After around 1 month in the atmosphere it is eventually oxidised to carbon dioxide (CO_2).

problems
It is readily absorbed into the bloodstream via the lungs, where it reacts with haemoglobin in the blood and stops it carrying oxygen to your cells. Chronic exposure to CO can cause heart disease and damage to the nervous system. Long-term exposure to high levels of CO can lead to death from oxygen starvation. CO is particularly dangerous to unborn foetuses, and can reduce birth weight and increase foetal mortality.

The fact that it eventually oxidizes to carbon dioxide means that it adds to the greenhouse effect and global warming. The effects on animals are the same as for humans.

hydrocarbons

what are they?
They are chains of hydrogen and carbon atoms that are present in both petroleum fuels and biodiesel. They are the main source of the characteristic diesel smell.

sources
Emitted from vehicle exhausts due to the incomplete combustion of fuel. They can also evaporate from fuels, and, with heat, also from engine oil.

problems
Eye irritation and respiratory problems. They are a major factor in the formation of smog, and ground-level ozone.

particulates

what are they?
They are physical particles, usually of carbon, of various sizes. Particles larger than 10 microns in diameter are not such a problem for human health, as they tend to settle relatively quickly, and are not easily inhaled. Fine particles of less than 10 microns in diameter (PM_{10}), however, remain in the air for longer, and can be inhaled deep into the lungs.

sources
Humans have been putting particulates into the air for millions of years, since they began to use fire, and there are natural sources such as forest fires and volcanoes.

Nowadays, the principal source of airborne fine particles is road traffic emissions, particularly from diesel vehicles, and they are highly concentrated in urban areas. In diesel combustion, hot exhaust gases moving into a cooler exhaust pipe can lead to spontaneous nucleation of carbon particles which are then emitted to the atmosphere.

problems
As well as creating dirt, odour and visibility problems, fine particulate matter is probably more dangerous to human health than any of the other chemicals mentioned here, even though we have been creating them for hundreds of thousands of years.

The smaller the particles, the more damage they can cause, as they can be breathed deeper into the lungs. Recent research has highlighted the dangers of particulates, including the possibility of asthma, bronchitis, emphysema, other lung disease and death. They can also transport carcinogenic compounds that they have picked up into the lungs.

volatile organic compounds (VOCs)

what are they?
Organic chemicals are those containing carbon, and are the basic chemicals in all living things. Most occur naturally, but some are produced synthetically. Many synthetic organic compounds are volatile – that is, they evaporate easily, and so can contribute to air pollution and cause health and environmental problems. Petroleum fuels are themselves VOCs, but the most important VOCs associated with vehicle fuels are benzene and 1,3-butadeine. Benzene is emitted from petrol vehicles but not diesel, and 1,3-butadeine is emitted from both.

sources
VOCs are emitted from vehicle exhausts, as unburnt fuel, and also from the evaporation of fuels when filling fuel tanks, and from spills etc.

problems
Benzene and 1,3-butadeine are known carcinogens, but only 1,3-butadeine concerns us here, as it is emitted from Diesel engines. As well as being carcinogenic, it can also cause damage to the central nervous system, liver and kidney damage, reproductive disorders and birth defects.

Other VOCs (hydrocarbons) play a role in the formation of ozone.

toxic organic micropollutants (TOMPs)

what are they?
They are a complex range of chemical compounds, emitted in very small quantities, and they include the polycyclic aromatic hydrocarbons, or PAHs (a range of carbon/hydrogen compounds ending in -ene), polychlorinated biphenyls, dioxins and furans.

sources

There are many types of TOMPs, but the ones that can be found in vehicle emissions are PAHs, dioxins and furans. They are produced by the incomplete combustion of fuels.

problems

Although they are emitted in very small quantities, they punch above their weight, and are carcinogenic, and highly toxic to wildlife. Their effect is enhanced by the fact that they accumulate and become concentrated higher up the food chain.

lead and heavy metals

Lead is a cumulative poison to the central nervous system, and can retard the mental development of children. As such, lead has been recognised as a very dangerous pollutant, and steps have been taken to reduce emissions.

Most lead in the atmosphere has come from vehicle emissions, as it has been used for many years as a petrol additive to protect engines from wear. Nowadays though, the introduction of lead-free petrol has meant a reduction in emissions of lead from petrol vehicles.

Diesel engines run on either mineral diesel or biodiesel do not emit lead.

secondary pollution problems

acid deposition

what is it?

It is either wet or dry precipitation of acids into the environment. If acids are deposited wet, it is commonly known as acid rain. Natural precipitation is slightly acidic, due to carbon dioxide in the atmosphere; rain is considered acid rain if its pH is less than 5.6. In some parts of the Eastern USA, rain pH of 2.3 has been measured (1000 times more acidic than natural rainfall).

sources

Mainly from fossil fuel emissions. Sulphur dioxide and nitrogen oxides are oxidized to sulphuric acid and nitric acid respectively.

Once oxidized, acids can be dispersed and precipitated very far from the emissions of the primary pollutants

problems
Soils and water bodies become more acidic, which damages plants and wildlife, as well as agriculture, forestry and fisheries. The effect is magnified in higher latitudes, as acids build up in snow during the winter, and are then deposited into water bodies in higher concentrations in the spring. Even if adult fish can survive such concentrations, their eggs usually can't.

Acids can leach nutrients from soils and plant tissues, restricting plant growth. Many soil micro-organisms can't survive a pH of less than 6, so acid precipitation kills them, reducing decomposition and the formation of healthy soils.

There has been extensive forest damage and die-back of trees in many locations in Europe and North America because of acid rain. Buildings suffer too – many features of historical buildings such as cathedrals (especially those built from limestone) have been eroded.

Finally, acid rain can affect the health of humans directly, especially causing allergies and colds in children; and via the leaching from soils of toxic metals such as aluminium and mercury, which can end up in drinking water, crops and fish.

ozone

what is it?
It consists entirely of oxygen, but instead of the usual O_2 molecules, ozone is in the form O_3. Ozone is considered a pollutant at ground-level only. High in the stratosphere, around 15-30km above the earth's surface, it does a useful job, shielding the earth from damaging ultra-violet radiation from the sun. About 90% of atmospheric ozone is in the stratosphere. It is the 10% at ground level that causes problems for humans.

sources
Ground-level ozone is not emitted directly; it is a secondary pollutant formed by sunlight, causing nitrogen oxides and hydrocarbons to react together.

It takes time for ozone to be formed, because of the two main nitrogen oxides, only one forms ozone; the other one destroys it! The initial nitrogen oxide emitted from vehicles, nitric oxide (NO) breaks up ozone into oxygen. However, as mentioned above, the longer the NO stays in the atmosphere, the more of it is oxidized to nitrogen dioxide (NO_2). NO_2 helps to form ozone.

For this reason, more ozone forms in rural areas than urban areas, unless air movements keep the nitrogen oxides in the city. So on hot, still, summer days where hydrocarbons and nitrogen oxides have built up from traffic, and NO_2 has had time to oxidize from NO, ozone build-up can be a serious problem.

problems
Ground-level ozone can be toxic to plants, including crops.
It can cause irritation to eyes, breathing difficulties, and ultimately lung damage.

As it is very reactive, it can also cause damage to fabrics, rubber and other materials

comparison of petrodiesel / biodiesel emissions

Most emissions are reduced by using biodiesel. Nitrogen oxides are the major exception and we will deal with these first. It is worth mentioning that a catalytic converter will in many cases make a bigger difference to emissions than the change to biodiesel alone. The two technologies work well together though: catalytic converters need zero-sulphur fuel to work well. Only the newest Diesels have a cat fitted as standard: it may be possible to retro-fit a cat to an older vehicle though this is usually only done for larger vehicles like trucks.

nitrogen oxides (NOx)

Most studies show that in a normal Diesel engine biodiesel can increase nitrogen oxides by up to 10%. It is possible for biodiesel use to actually decrease these emissions, but only if the engine timing is retarded. This process, which can be done by a competent mechanic, matches the timing of the fuel injection to the combustion characteristics of vegetable oil. All things being equal this step reduces NOx emissions without increasing other emissions. Even though it will cause a slight performance penalty it seems theoretically to be a good idea.

For a number of reasons, however, it is not always the best option. Firstly, retarding the timing will mean that petrodiesel is less efficiently burnt, meaning that the engine will produce greater emissions if burning petrodiesel. Many vehicle users will not be running on 100% biodiesel all the time, especially if they are on long journeys or choose to blend petrodiesel with biodiesel for easier starting in winter. Secondly if the vehicle is fitted with a catalytic converter, retarding the timing will reduce its effectiveness overall, increasing emissions of other pollutants.

Figures given below reflect this effect. In general it seems that changing the timing is most appropriate for vehicles without cats which will only rarely use petrodiesel. The rest of the time we have to accept that NOx emissions from biodiesel will be slightly higher than those for petrodiesel.

NOx emissions compared to low-sulphur diesel without catalytic converter:

- biodiesel   +6%
- biodiesel with catalytic converter   +5%
- biodiesel with catalytic converter and timing change -30%

sulphur dioxide (SO₂)

SO2 emissions are virtually eliminated by using biodiesel

carbon monoxide (CO)

CO emissions compared to low-sulphur diesel without catalytic converter:

- biodiesel   -15%
- biodiesel with catalytic converter   -98%
- biodiesel with catalytic converter and timing change -94%

hydrocarbons

Hydrocarbon emissions compared to low-sulphur diesel without catalytic converter:

- biodiesel   -38%
- biodiesel with catalytic converter   -92%
- biodiesel with catalytic converter and timing change   -86%

particulates

There is a noticeable reduction in the visible black soot (carbon) particulates emitted, and in finer particles too. A LILI course participant who had been running his vehicle on biodiesel took his car for its MOT, and the smoke test didn't register particulates at all. The mechanic thought that his machine was broken.

Particulate traps are available and fitted as standard to some buses and trucks. These can dramatically reduce particulates from both petrodiesel and biodiesel.

Particulate emissions compared to low-sulphur diesel without catalytic converter:

- biodiesel   -31%
- biodiesel with catalytic converter   -68%
- biodiesel with catalytic converter and timing change   -50%

volatile organic compounds (VOCs)

If biodiesel is made correctly, with methanol recovery and washing, VOC emissions will be reduced by around 50%

toxic organic micropollutants

Tests in the US have recorded a 74% reduction in polyaromatic hydrocarbons from the use of 100% biodiesel.

lead

Neither mineral diesel or biodiesel contain any lead.

acid deposition

Sulphur dioxide is the main contributor to acid rain; as we have seen, biodiesel doesn't contain any sulphur, or emit any sulphur dioxide. Nitrogen oxides are minor contributors, and these emissions may increase with biodiesel use. Catalytic converters and / or timing changes should bring these to the same levels as mineral diesel, or less.

ozone

We have seen that ozone is formed by NO_2 reacting with hydrocarbons in sunlight. As well as hydrocarbon emissions being almost completely removed (with a catalytic converter), the ozone forming potential of hydrocarbons emitted by burning biodiesel is half of that for mineral diesel.

other environmental benefits

positive energy balance

Mineral diesel has a negative energy balance of around 0.8:1 That is, it takes a bit more energy to bring it to the point of use than is released when it is burnt. In that case you may wonder why they bother! Well the answer is that people are willing to pay the economic cost of all that extra energy in order to get liquid energy in a usable form where they want it. Paying for the environmental costs of all that extra energy (which is of course also fossil fuel energy) is a different matter. However, multiple sources indicate an energy balance for biodiesel of around 2.5:1. That is, a given amount of biodiesel will provide 2.5 times as much energy when burnt, as the energy that was used to produce it. This is the best ratio of any road fuel.

So biodiesel not only saves fossil fuel resources (and the carbon dioxide and other emissions that go with them) because it is itself renewable, but also because it uses much less fossil fuel in its production and distribution.

The reasons are obvious when you think about it. Compared to the effort required to set up oil extraction plants in far flung corners of the globe, often in very harsh conditions, drill into the earth's crust for crude oil, then fractionally distil it into diesel fuel, and transport it around the earth in supertankers, planting, harvesting and pressing seeds is very small beer indeed. Most of the effort in ripening the seeds is carried out by the sun, for free. Note that these figures apply to biodiesel made from rapeseed oil grown using intensive agriculture (with heavy mechanisation and high levels of fertiliser input).[1] The energy balance for biodiesel made from waste vegetable oils will be higher: 6:1 is a typical figure.

energy density

Energy density is a measure of how much energy is in a given volume of fuel. High energy density fuels are good because they require less energy to transport. Biodiesel has a similar (slightly lower) energy density to mineral diesel. This means that it contains more energy per kilogram than other biofuels, and so less energy is needed to transport each unit of energy. A drawback of some other biofuels is that they are not available in liquid form and so have a very low energy density.

biodegradable

Mineral diesel has a much more complex chemical structure, and is therefore much more difficult to biodegrade than biodiesel.

Spills of biodiesel into rivers or local environments can occur, and although this can cause environmental damage, biodiesel biodegrades rapidly, and will be absorbed relatively quickly. Tests have shown that biodiesel will have biodegraded almost 90% in one month, which is about the same as sugar, and four times faster than mineral diesel. Blends will degrade faster than mineral diesel, and the rate of increase is higher than the ratio of biodiesel in the blend (e.g. a blend with 20% biodiesel will degrade twice as fast as mineral diesel).

local

There are enormous environmental benefits associated with local-scale production and consumption, and the benefits increase the more small-scale producers and consumers there are. Small-scale producers will collect locally, from local food outlets (waste oil) or farmers (fresh oil) and produce biodiesel for themselves, and perhaps a few local customers. This means shorter distances for waste oil, fresh oil and biodiesel to be transported, and if there are any spillages, smaller amounts will be involved, which can be absorbed by the environment more easily.

---

[1] Imperial College Centre for Energy Policy and Technology
www.iccept.ic.ac.uk/

It also helps to ease the pressure from those in the oil industry eager to drill in national parks and wilderness areas such as Alaska.

spills

If biodiesel is not shipped across the world's oceans, then it won't be spilled onto the world's coastlines when the inevitable accidents occur.

non-toxic for environment

Tests in California in the 1990s on larval stages of fish and shrimps (used because they are more susceptible to toxicity than adults) found that concentrations of biodiesel and mineral diesel required to kill 50% of test samples were 578 ppm (parts per million) and 27ppm respectively for fish, and 122ppm and 2.9ppm for shrimps.

utilizing a waste product

It's very difficult to ascertain exactly how much waste vegetable oil is produced per year in the UK, but it is at least 100,000 tonnes, and could be as much as 500,000 tonnes. Much of this is disposed of illegally – either down drains, or into landfill via the refuse system. Waste oil for pig feed is illegal, as is cattle feed from October 2004, because of EU legislation. Because of the BSE crisis, the UK should have been in the vanguard of compliance with this legislation, but we applied for a two-year deferment of the ban, which expired on October 31 2004. (To be fair, the deferment was to ensure that some other industry – i.e. biodiesel – could be developed to take this oil, so that it didn't end up in sewers or landfill).

So the 20p reduction in duty was introduced. This was carefully calculated to encourage a fledgling biodiesel industry just enough to soak up the extra waste oil that would come online when the cattle feed ban was enforced. Biodiesel production can utilize a waste product.

The biodiesel v. straight oil debate as regards the environmental benefits is a complicated one. The use of straight vegetable oil negates the problem of using toxic chemicals and energy in the biodiesel production process. However, your car may be old, and not worth a conversion to run on straight oil. It is more environmentally beneficial to continue to run an old car than to buy a new, cleaner one (more environmental damage is caused in the production of a car than in running it during its lifetime). When all new diesel vehicles are able to accept vegetable oil, then this may be the greener option.

production by-products

The production of biodiesel causes 79% less wastewater, and a reduction of 96% in hazardous solid waste than the production of mineral diesel (US dept of energy, 1998). These are little-known issues, but of enormous benefit to the environment.

but biodiesel isn't *necessarily* green

As promoters of biodiesel we have to be very careful to point out that biodiesel is not a catch-all solution to the environmental and ethical problems of petrodiesel. Of course it takes more than a change of fuel to make a whole transport system green. In fact some environmentalists criticise the biofuels movement for giving drivers a 'get-out clause'. We think it's important to approach the issues from both sides at once: more public transport **and** better fuels.

An even more difficult issue is the fact that biodiesel is now being made just to make money as petrodiesel goes up in price, ignoring the green possibilities of the fuel. Some of the biodiesel sold in the UK at the moment is made in Germany from rapeseed grown intensively in Scotland. This pretty much wipes out the energy balance advantages we have explored. An even less appealing scenario is the growing of GM oil-bearing crops in Brazil by Monsanto to make biodiesel that will then be shipped to the UK, something that is already in the pipeline. We can expect to see people being deprived of access to land for subsistence needs in some of the poorest countries of the world, in order to fuel vehicles in richer nations. Could it be that the global biodiesel economy has the potential to be even worse than the oil economy?

Biodiesel does have the potential to be a significant part of a greener, saner world. We have to be careful when promoting this vision not to end up giving support to those who would use the good image of biofuels to market their environmentally-damaging product.

health benefits

emissions

Some of the human health problems associated with the burning of fossil fuels are outlined in the 'emissions of pollutants' section.

Below is a brief summary:
Sulphur dioxide in ambient air is associated with asthma and chronic bronchitis; it is almost totally eliminated by switching to biodiesel.

Carbon monoxide is a poisonous gas which restricts the oxygen-carrying capacity of the blood, and can lead to death by oxygen starvation, and increase foetal mortality. It is reduced dramatically by using biodiesel, especially with a catalytic converter.

Burning fossil fuels releases a range of aromatic hydrocarbons known to be carcinogenic, and to damage human immune, reproductive and hormone systems. Burning biodiesel instead of mineral diesel reduces these compounds by 50% to almost 100% in some cases.

Tests in the States have found that particulates released on burning diesel is reduced by a third by switching to biodiesel, and by more than half when used in conjunction with a catalytic converter. Particulates are carcinogenic and cause respiratory disease.

Ground-level ozone and smog can cause a range of respiratory problems. Emissions of hydrocarbons (a major factor in ozone and smog production) are in the range of 50-100% lower with biodiesel; plus the ozone forming potential of hydrocarbons emitted by burning biodiesel is half that for mineral diesel.

toxicity

Properly made biodiesel is essentially non-toxic (10 times less toxic than ordinary table salt) and is not a skin irritant.

In a promotional video, the director of Global Commodities drinks some of their biodiesel to show just how non-toxic their product is. We wouldn't advise this, but he's still alive[2].

safety

Biodiesel has a much higher flash point than mineral diesel. Flash point is the minimum temperature at which fumes can ignite in the air above the fuel (if there is a source of ignition). The flash point of mineral diesel is about 60°C, and for biodiesel it is around 150°C. By blending biodiesel with mineral diesel you can increase its flash point.

benefits for the vehicle

lubricity

Biodiesel is much more lubricating than mineral diesel. In diesel fuel systems, pumps are lubricated or partially lubricated by the fuel. So higher fuel lubricity can mean lower wear and longer life for the pump.

The sulphur that naturally occurs in crude oil was left in mineral diesel to improve lubricity. When the environmental and health problems of sulphur emissions were understood, ultra-low sulphur diesel was introduced, with a 3p per litre duty cut in the UK. This meant that people moved to it en masse. Now almost all petrodiesel at the pumps is ultra-low sulphur, which will be mandatory from 2005. There are new laws (especially in the US) to drastically reduce sulphur emissions, and sulphur-free fuels will be mandatory in the UK from 2009.

Removing the sulphur has resulted in lower lubricity in mineral diesel, which could be a problem (premature wear, leading to breakdown / shorter engine life). Not for biodiesel though, which has higher lubricity without the sulphur. Even blends involving tiny amounts of biodiesel will dramatically improve lubricity. There is an incredible figure on the website of the National Biodiesel Board in the US: a blend with 1% biodiesel will increase the lubricity of mineral diesel by 65%. Diesel engine manufacturers have pointed out that not only is it a lubricating additive, it is also a fuel, and so

---

[2] At the time of writing

this gets round the problem of other lubricating additives reducing the efficiency of the fuel. They say that just 2% biodiesel added to mineral diesel is enough to solve the lubricity problems associated with low-sulphur diesel, but if more than 2% is used, it causes no problems because it is a fuel rather than just an additive.

Garages reported a big increase in problems with fuel pumps on trucks after the amount of sulphur was reduced in diesel. This is one of the reasons the French introduced 5% biodiesel into all their diesel at the pumps; they didn't trumpet the environmental benefits of their decision, as the primary reason was to help lubricate their trucks better.

There is an audible difference when biodiesel is used. Biodiesel users consistently report a quieter engine, and much more of a purr.

cetane number

Cetane number is to diesel fuel what octane number is to petrol. It is a measure of the fuel's ignitability. Cetane is a hydrocarbon molecule that ignites very easily under compression, so it was assigned a rating of 100. All the hydrocarbons in diesel fuel are indexed to cetane as to how well they ignite under compression[3]. The higher the cetane number, the more ignitable the fuel.

Biodiesel generally has a higher cetane number than mineral diesel. It varies based on feedstock (the oil from which the biodiesel was made). The cetane number of biodiesel depends on the distribution of fatty acids in the original oil or fat from which it was produced. The longer the fatty acid chains (see 'the chemistry') and the more saturated the molecules, the higher the cetane number.

Biodiesel does not need the cetane improver additives that are added to 'premium' petrodiesel.

other benefits

Energy security: biodiesel can reduce countries' dependence on remote supplies of fuel, especially from politically volatile areas of the world. Ultimately, switching to renewables could help reduce

---

[3] http://encyclopedia.thefreedictionary.com/Cetane+rating

the potential for military conflict as powerful interests jostle for dwindling fossil fuel reserves.

It can be used with existing infrastructure: we don't have to replace the world's Diesel engines or retrain mechanics, representing a massive saving in time, money and resources over some other fuel options.

Biodiesel production can help to increase farm income, which in turn might help to reduce the subsidies paid to farmers in the developed world (these have been criticised because they give western farmers an unfair advantage in global markets over farmers in developing countries).

Distribution of wealth and power: the growing of fuel crops can help put fuel supply into a greater number of hands, down to the local scale, and even to individuals.

the Diesel engine

Rudolf Diesel

Rudolf Diesel (1858-1913)

"The use of plant oil as fuel may seem insignificant today. But such products can in time become just as important as kerosene and these coal-tar-products of today."

Rudolf Diesel in the year 1912, in his application for a patent.

The Diesel engine was designed by Rudolf Diesel in 1892. The first working engine was built at the Augsburg Maschinenfabrik (now part of the MAN B&W group) in 1897. The single cylinder engine, weighing five tonnes, was used to power stationary machinery. It had a single 10 ft (3 m) iron cylinder with a flywheel at its base, and produced 20 hp at 172 rpm! The engine operated at 26% efficiency, a very significant improvement on the 20% achieved by the best petrol engines of the time.

On February 27, 1892, Diesel filed for a patent at the Imperial Patent Office in Germany. Within a year, he was granted Patent No. 67207 for a "Working Method and Design for Combustion Engines: a new efficient, thermal engine." With contracts from Frederick Krupp and other machine manufacturers, Diesel began experimenting and building working models of his engine. In 1893, the

The first Diesel engine.

first model ran under its own power with 26% efficiency, remarkably more than double the efficiency of the steam engines of his day. Finally, in February of 1897, he ran the "first Diesel engine suitable for practical use", which operated at an unbelievable efficiency of 75%.

Rudolf Diesel literally disappeared in 1913. There is some question over the cause of Diesel's death. Diesel did not agree with the politics of Germany and was reluctant to see his engine used by their naval fleet. With his political support directed towards France and Britain, he was on his way to England, possibly to arrange for them to use his engine, when he inexplicably disappeared over the side of the ship in the English Channel. Whether by accident, suicide or at the hand of others, the world had lost a brilliant engineer and biofuel visionary.

development of the Diesel engine

The 1920s brought a new injection pump design, allowing the metering of fuel as it entered the engine without the need for pressurised air and its accompanying tank. The engine was now small enough to be used in vehicles. 1923-1924 saw the first lorries built and shown at the Berlin Motor Fair. In 1936, Mercedes Benz built the first automobile with a Diesel engine, theType 260D.

Mercedes Benz 260D.

the modern Diesel engine

The modern Diesel engine works in essentially the same way as an engine from the late 1930s, though there have been many advancements in manufacture and engine efficiency.

A Diesel compresses the air in the cylinder at a ratio of about 22:1. When the atomised fuel is injected into the cylinder it ignites because of the high temperature the air in the cylinder reaches as it is compressed. Petrol engines have a lower compression ratio of about 9:1. A spark from the spark plug ignites the petrol that is atomised in the cylinder. This is the main difference between petrol and diesel-fuelled engines.

It is necessary to heat up the air in the cylinder to initiate the cycle. This is done with a glow plug that is deactivated once the engine has started.

the Diesel cycle

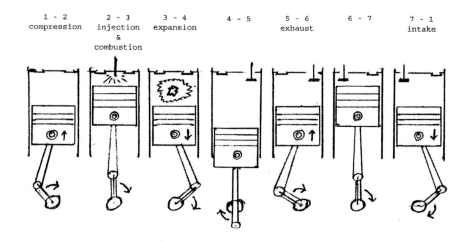

The cycle of a compression-ignition (Diesel) engine

In this diagram of the cycle of the Diesel engine:

1   the air is compressed in the cylinder by the rising piston
2   the fuel is injected and ignites under compression
3   The fuel explodes, forcing the piston down, which propels the crankshaft round
4   The crankshaft is connected to the camshaft which in turn opens the exhaust valve
5   The rising piston forces the exhaust gases out of the exhaust valve
6   The camshaft opens the air intake valve
7   The falling piston sucks new air into the cylinder to start the cycle again.

fuel injection

The major differences between modern Diesel engines are in the way fuel is injected into the cylinder

indirect injection (IDI)

In an IDI engine the fuel is injected into a pre-chamber or 'swirl chamber' before passing into the cylinder. The connecting passage and space can be designed to produce good swirling and mixing of the fuel and air, hence better combustion. This type of control usually allows for a lower overall fuel-to-air ratio and lower pollutant emissions. Until recently it was not feasible to make small direct injection engines, so most Diesel cars over ten years old are IDI. Some of these engines have the ability to use SVO but the fuel system is usually the limiting factor.

direct Injection (DI)

In a DI engine the fuel is directly injected into the combustion chamber, directly above the piston. This design is also referred to as an open chamber engine. The larger of these engines are to be found in ships, trains and tractors amongst others. They usually work at about 1500 rpm. The fuel is dispersed via a set of 8-10 holes in the injectors into a fine mist. The air content of this mist contains enough oxygen to combust. These large DI engines tend to be very fuel tolerant and many will run on SVO.

For many years technical difficulties meant that Direct Injection engines had to be large and heavy. In the late 1960s, largely through the work of Elsbett, a German engineer also heavily associated with biofuels research, smaller direct injection engines began to be feasible. Since the mid 1990s this has become the dominant type of Diesel engine in vehicles. (TDI, SDI, HDI etc.). Unfortunately DI engines tend to be less tolerant of differences in fuel viscosity than IDI engines.

fuel systems

There are a number of different fuel systems on modern Diesel engines. They differ in the way the fuel pump and injectors are connected and controlled.

mechanically-controlled fuel injection

Fuel is supplied to the injector by the fuel pump, which is mechanically driven from the crankshaft. The drive from the crankshaft is set so that the pump delivers fuel at the correct time in the engine operation cycle. There is a separate fuel pipe from the pump to each injector. Most Diesel engines more than ten years old use a system like this. There are two main types:

- **in-line injector pump**
  In-line pumps have the pipes coming out of the top. These pumps have a separate internal pump for each injector pipe. They are typically better at dealing with viscous fuels. These pumps are only common on larger and more 'agricultural' vehicles

- **rotary injector pump**
  Rotary pumps are similar in appearance to a petrol engine distributor. A single pumping mechanism rotates and supplies fuel to each cylinder in turn. These vary in their ability to deal with viscous fuels like SVO. As a rule of thumb, injector pumps made by Lucas and CAV tend to be less tolerant of high-viscosity fuels than those made by Bosch or most of the Japanese manufacturers.

electronically-controlled fuel injection

Recently developed electronically-controlled injectors can provide exact amounts of fuel at very high pressure, very precise timing and even multiple injections within each cycle to give greatly improved combustion, and in turn increased fuel economy, lower engine noise and cleaner emissions. By monitoring the engine using a number of sensors, the electronic controller can modify the fuel injection characteristics to improve combustion. These systems are typically less fuel tolerant than mechanical systems. They are also dependent on fuel sensors, which can easily be damaged by methanol and perhaps other solvents.

- **common rail injection systems (CDI)**
  With this system a pump constantly supplies fuel at a very high pressure to the common rail - a tube with thick walls. From the common rail, fuel is supplied to electronically-controlled injectors. The higher pressure injection gives a finer spray and improved combustion.

- **unit injector system and unit pump system**
  Unit injector systems combine the pump and injector into one unit. The pump is driven from the engines camshaft. Fuel delivery is timed and metered by electronically-controlled valves. Unit pump systems are similar with the pump and injector separated by a short high-pressure line. Some VW Golf TDIS use this sort of system.

fuel pipes, filters and seals

There are a number of links in the chain from fuel tank to cylinder. Typically the fuel travels down a pipe under the car to the engine bay. It then passes through the fuel filter that removes particulate matter and water, sometimes assisted by a small and simple lift pump. It then travels to the fuel pump and up to the injectors. Between each point there are pipes that may be made of mild steel, copper, rubber or plastics. If you are using biodiesel for a long period, rubber hoses will fail, as they are not compatible with biodiesel. It is relatively easy to find and replace these hoses with Viton or similar fluoroelastomer hoses. Vehicles made since 1995 will almost certainly have this type of hose already. The lift pump and fuel pump contain seals that may also be made of rubber on older vehicles. This could potentially create problems, as especially in the fuel pump these seals will be hard to replace. However we know of many people who are using biodiesel on older vehicles without problems of this nature.

turbocharging

Many diesels are fitted with a turbocharger. This takes the energy in the exhaust and uses it to compress the air going into the engine, allowing for a higher efficiency of combustion. An intercooler is fitted to deal with the higher temperatures this causes. There should be no turbo issues with biodiesel or SVO.

issues with using biodiesel

mileage

It is generally reported that using biodiesel can reduce range (the distance you can travel on a full tank) by up to 3%. However, our course participants who have been using 100% biodiesel tell us that they get more miles per gallon (or should we say kilometres per litre?). It probably depends on the feedstock, the biodiesel-making method, and the engine. It seems that it's a pretty small effect in any case. Using biodiesel in blends with petrodiesel can increase the efficiency of combustion of the whole mix. Maybe that's why diesel from pumps in France always seems to last forever!

cold starting

Petrodiesel is at the heavier end of the spectrum of fuels obtained from crude petroleum oil, and so there is a problem that in very cold weather, it will start to gel at higher temperatures than petrol. There are several ways to avoid this problem. The obvious one is to park your diesel vehicle in a garage, but you could also add a little kerosene, or other special cold start additives (winterising agents), or have some sort of heating element in the fuel tank, fuel filter or somewhere on the fuel line. Many of the SVO conversion kits use equipment that was designed for petrodiesel in the cold.
In the UK, we don't experience the kind of temperature extremes found in North America or continental Europe, and so there is rarely a problem starting petrodiesel vehicles throughout the winter. Petrodiesel bought in winter is usually pre-blended with winterising agents.

Depending on feedstock, biodiesel can be similar in cold-weather performance to petrodiesel or a lot worse. Feedstock that is highly saturated makes biodiesel that can turn solid above 0ºC. There are a number of solutions to this problem. You could make 'summer' and 'winter' biodiesel with different grades of WVO. Many people just add petrodiesel if they are experiencing starting difficulties.
Trials in the US have found B20 (a blend with 20% biodiesel) to have the same or better cold starting properties as mineral diesel, and that this blend worked down to temperatures of −32ºC. You could also decide to leave methanol in the fuel if you are sure your engine can handle it.

Biodiesel produced for sale generally contains winterising agents (as does petrodiesel sold in cold countries). Suppliers are listed in the resources section.

Two terms are used in the literature regarding starting in extremely cold temperatures – the first is cloud point, which means the temperature at which small solid particles are first seen to gel as the fuel is cooled. The second is cold filter plugging point, which means the temperature at which the fuel filter becomes blocked due to the accumulation of solids in the fuel.

biodiesel acting as a solvent

Biodiesel is an excellent solvent, and so could potentially deposit materials picked up from storage containers if they are not completely clean. Also, deposits may build up from mineral diesel in your fuel tank. Biodiesel could possibly remove them and deposit them in the fuel filter. It's a good idea to check your fuel filter reasonably regularly when first using biodiesel. It will probably need replacing fairly early on. If you continue to use biodiesel, it won't be a recurring problem.

Also, be sure to wipe biodiesel off any painted surfaces, because its solvent effect can mean that it could remove paint.

fuel quality

Quite a few years ago now industry was at last forced to remove the naturally-occurring sulphur from diesel for environmental reasons (the sulphur turned into sulphur dioxide in the engine, escaped into the atmosphere, and eventually came down as acid rain). The problem with this is that engines have been wearing far more quickly because of the lack of lubrication: the sulphur in the fuel increased the lubricity of the oil (like tetra-ethyl lead added to petrol). By using biodiesel in a blend you are not only improving the ignitability of the petrodiesel (increasing efficiency) - the increased lubricity helps reduce wear of the engine components.

There are issues around the quality of the biodiesel that you make. If the quality is too poor, then it may contain water if you do not de-water properly after water washing. This can cause pitting of cylinders, and eventually engine failure. Also, residual methanol can cause de-laminating of sensors in new cars. Any excess catalyst can cause deposits on the injectors and cylinders (coking).

Impurities, especially glycerine and unfiltered gums can cause injector coking which can lead to an uneven fuel spray, uneven heating of the cylinder, and possible engine failure. Straight vegetable oil used in a normal unmodified Diesel engine will certainly cause these deposits due to the glycerine content.

Luckily it is not hard to make biodiesel which is of sufficient quality to avoid these issues.

Biodiesel contains very slightly less energy per litre than mineral diesel, but you shouldn't notice any difference in the performance of your car, except maybe that it's quieter.

engine warranties

Most new cars are sold with a three-year warranty. If you have a new car the warranty may technically be invalid if you run on 100% biodiesel ('B100'). Volkswagen Group (VW, Seat, Audi and Skoda) warranty many of their Diesel vehicles to run on B100 as long as it meets EN14214. Most other manufacturers say B5 is fine. More detailed information on a model-by-model basis is available on the Channel 4 website (see 'where can I find out more?'). Not all manufacturers have tested biodiesel extensively: it tends to be those who sell into the German market, where biodiesel is more widely used, who have detailed compatibility information. The fact that the manufacturer does not warranty their engine to be used on B100 does not mean that it will not work. Part of Peugeot's own fleet has been running on B30 for some years even though they do not warranty the same models to run on the same blend.

For vehicles over 3 years old warranty is a non-issue, but it can be a problem when trying to persuade people to use biodiesel in vehicle fleets, where warranties are often in-date and important. We have not yet been able to find any example of test cases where warranty claims have been invalidated by use of biodiesel. It might be that the manufacturer passes the claim onto the manufacturer of the fuel system in such a case.

biodiesel chemistry primer

This is a working chemistry primer designed to give you the basic chemistry knowledge you need to understand what is happening when you make biodiesel. It has been written by a non-chemist for non-chemists. You can easily make biodiesel without understanding any of this, but it definitely helps to have some theoretical background.

the building blocks

Biodiesel chemistry is essentially organic chemistry, which is chemistry based around carbon. Carbon can combine with other elements in many ways to make different molecules, all of which have different properties.

The basis of organic chemistry is that elements can connect together to achieve a more stable configuration. Each element is trying to get to the point where it has 8 electrons in its outer shell. It can achieve this by sharing electrons with other elements. This sharing binds the elements together into a molecule. The connections formed between the elements are known as bonds.

To get an idea of how the elements combine, we can imagine elements as balls with different numbers of holes in them. For example, carbon is a ball with 4 holes, Oxygen has 2 holes, Hydrogen has 1 hole. If we connect the balls together with rods that slot into the holes, we can create simple or more complicated molecules. It's just like meccano! Reactions between chemicals tend to produce stable molecules. In our ball and stick model this means that all the holes are filled.

An example of a stable molecule is methane:

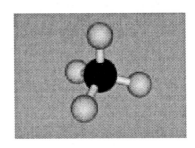

Carbon is the larger ball in the middle; the balls around the edge are hydrogen. Notice that carbon has 4 sticks attached to it, while each hydrogen has 1 stick attached to it. This structure is represented in 3-D to show more accurately how the molecule fills space, but is easier to draw in 2 dimensions:

$$
\begin{array}{c}
\text{H} \\
| \\
\text{H} - \text{C} - \text{H} \\
| \\
\text{H}
\end{array}
$$

This is the style we will use for the rest of the tutorial. When writing about common organic molecules chemists often don't bother to use pictures and will use text abbreviations. In this style, methane is known as CH_4 (carbon with 4 hydrogens). It's a less informative notation but it's easier to type!

There are a relatively small number of elements involved in basic biodiesel chemistry. Here are the most important ones

element	symbol	valency
Carbon	C	4
Hydrogen	H	1
Oxygen	O	2

Valency is the technical name for how many electrons the element has to share (how many holes in the ball).

With these simple elements as building blocks we can make many different molecules with different properties.

chains

The simplest types of larger molecules are hydrocarbon chains: chains made out of hydrogen and carbon only. Starting with methane we can make longer chains by connecting 2 or more carbons together:

alkanes

H—C—H (with H above and below)	H—C—C—H (with H above and below each C)	H—C—C—C—H (with H above and below each C)
methane	ethane	propane

These molecules all have similar but not identical properties. Together they are known as alkanes. The principle can be extended to make very long chains, which are known by the number of carbons in the chains. So a C20 alkane has 20 carbons in a chain and would be known as eicosane (eicos is Greek for 20). Its shorthand notation would be $C_{20}H_{42}$ and it would look like this:

Petrodiesel consists of a mixture of different long-chain alkanes with smaller amounts of other things. The average chain length is about C20.

In alkanes, the carbons are held together by one single bond (one shared electron). Because each carbon has 4 electrons to share, 2 carbons could also share 2 electrons between them and each one still have 2 electrons to share. This kind of bond between the carbons is known as a double bond and the simplest chains using double bonds are the alkenes:

alkenes

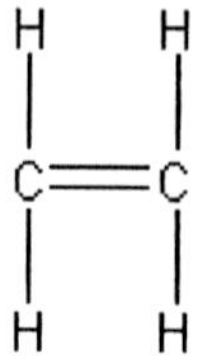

ethene 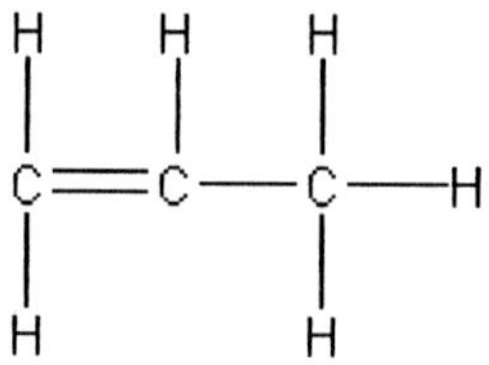   propene

Equally, the carbons could share 3 electrons and have one electron to share. This is known as a triple bond. The simplest chains using a triple bond are the alkynes:

alkynes

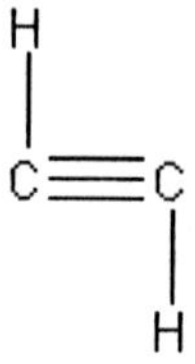

ethyne 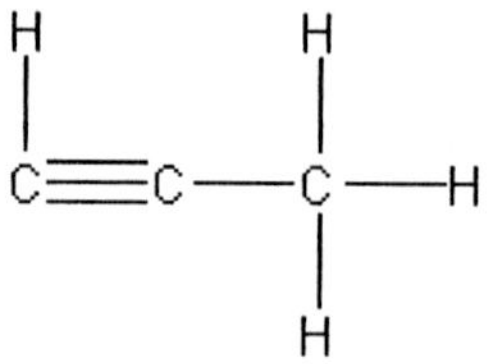   propyne

Compounds containing single bonds are known as **saturated**. Compounds containing double or triple bonds are known as **unsaturated.** Saturated compounds are more stable than unsaturated compounds: double or triple bonds can be broken allowing another bond to be made.

These hydrocarbon chains are a fundamental building block of organic chemistry and form the backbone of most of the compounds we are interested in.

functional groups

A functional group is an atom or group of atoms in an organic compound that gives the compound some of its characteristic properties. A typical organic molecule consists of a hydrocarbon chain with one or more functional groups attached. The characteristics of the overall molecule depend more on the functional group than on the length of the chain. One on the simplest functional groups is the hydroxyl group OH. When this is attached to different length chains it forms a series known as the alcohols:

alcohols

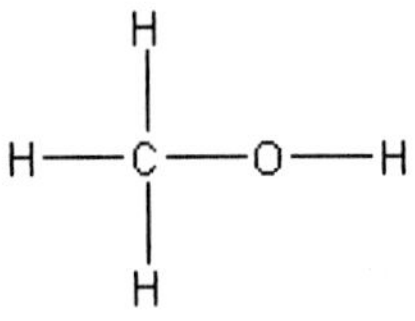

| methanol | 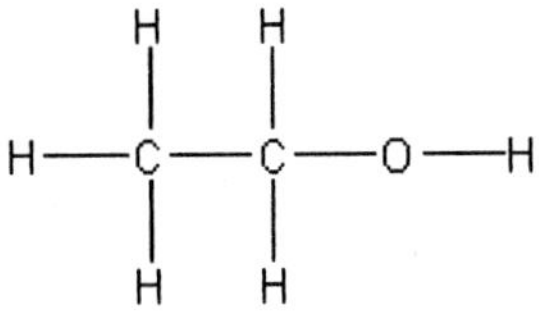 ethanol | 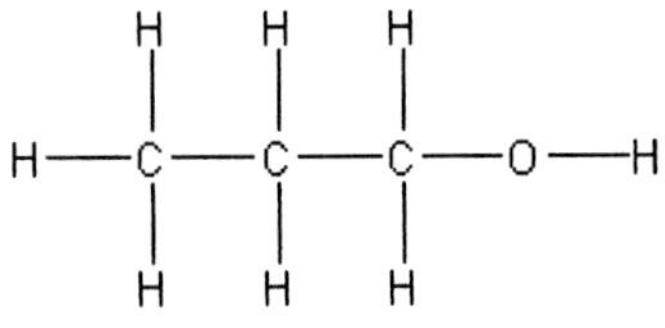 propanol |

The alcohols all have similar properties because they all have the same functional group: the hydroxyl group (OH).

This is an important functional group in biodiesel chemistry. The other important group is the carbonyl group COOH containing a double bond between carbon and oxygen. This gives rise to two important series.

fatty acids

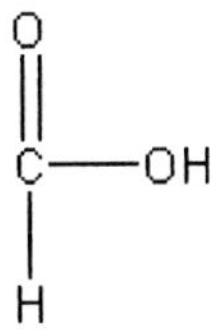

| methanoic acid | 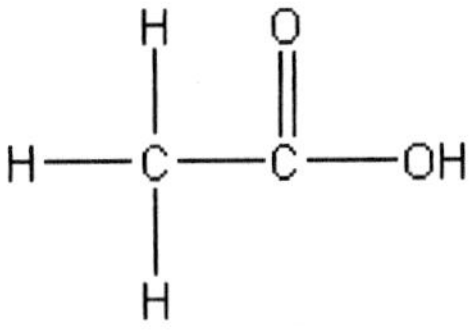 ethanoic acid | 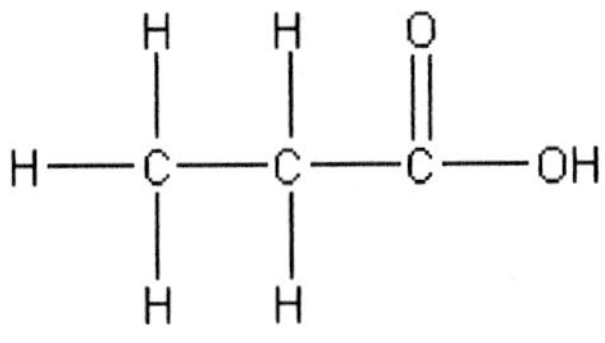 propanoic acid |

Fatty acids terminate in a hydrogen attached to the single-bonded oxygen (i.e. the one on the right in these diagrams). The other side of the carbonyl group is attached to a hydrocarbon chain except in the simplest fatty acid (methanoic acid). Used vegetable oils contain fatty acids.

esters

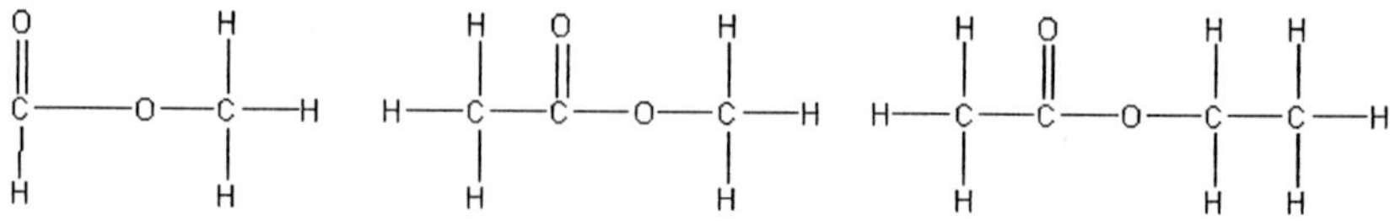

methyl methanoate   methyl ethanoate   ethyl ethanoate

Esters have a hydrocarbon chain attached to the single-bonded oxygen. The other side of the carbonyl group is also attached to a hydrocarbon chain except in the simplest ester (methyl methanoate). The first two esters above are called methyl esters because they have a methyl group (CH_3) attached to the single-bonded oxygen.

Esters can be formed from fatty acids and alcohols by a process known as **esterification**.

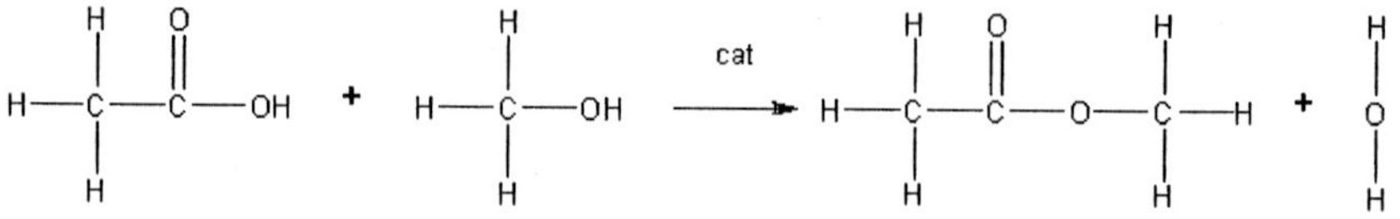

ethanoic acid   methanol   methyl ethanoate   water

This process needs a catalyst to take place, usually a concentrated acid.

A useful way of thinking of an ester is as an alcohol attached to a fatty acid.

Both vegetable oil and biodiesel are largely esters. The biodiesel we make is a type of methyl ester.

putting it all together

The worst part is over! We have now got the basic knowledge to understand the structure of vegetable oil and the process of transesterification. The next diagram shows a vegetable oil molecule or triglyceride. It looks rather complicated but by breaking it down into recognisable parts we will make more sense of it.

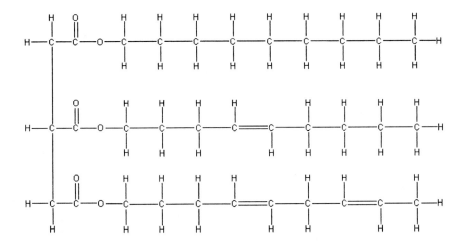

The first thing we should notice is the functional group. It is the carbonyl group, and we can see that this molecule is an ester. In fact it's a triester: there are three places where the carbonyl group's single-bonded oxygen is connected to a hydrocarbon chain. So what about those chains? We can see that some are saturated whereas others contain one or more double bonds. In fact the second chain is mono-unsaturated and the third chain is poly-unsaturated. In a real-life triglyceride the chains might be of different lengths depending on the type of vegetable oil. They would also vary in their degree of saturation: palm oil chains would be more saturated than sunflower oil chains for example. The fact that the chains can be quite different without changing the basic chemistry means we can simplify the diagram by representing them with the shorthand term 'R'. 'R' means a hydrocarbon chain. We can number them R1 R2 R3 to show they are different.

That already looks a lot more manageable. One thing we can still see is that the molecule is rather big: in fact it's a lot bigger than the esters we looked at before. The size of the molecule is what gives this triglyceride its viscosity, and that viscosity is, as we have already seen, the major problem facing us if we want to use vegetable oil as a fuel in conventional Diesel engines.

As we saw before, esters can be thought of as an alcohol attached to a fatty acid. In this case we have 3 different fatty acids attached to a large alcohol molecule. The alcohol in question is glycerol, which is a triol (it has 3 OH functional groups).

glycerol

We can see that this glycerol backbone holding together the three fatty acids is the key to the large size of the molecule. What we want to do is replace this with a smaller alcohol.

transesterification

This brings us to the heart of biodiesel chemistry: transforming the vegetable oil esters into the biodiesel esters. This is known as **transesterification**, turning one ester (glyceryl esters) into another ester (methyl esters). We do this by adding methanol to the triglyceride. This removes the glycerol backbone and replaces it with a smaller methyl group, thus splitting apart the large molecule .We are left with methyl esters and glycerol as separate products.

triglyceride   methanol   methyl esters   glycerol

Transesterification is a reversible reaction. This means that as the reaction progresses from left to right creating methyl esters and glycerol the opposite is also happening: some of the products are re-combining to form trigylcerides. At some point the reaction will reach equilibrium where there is a certain proportion of unreacted vegetable oil to biodiesel. We want to make sure that the reaction proceeds as far as possible, which can be achieved in two main ways. The first is to remove the products as they are formed. In large-scale industrial plants this is done by removing the glycerol using separators. This is impractical in smaller set-ups. Instead we use an excess of methanol to 'push' the reaction towards completion; there's always enough methanol for the left-to-right reaction to happen more readily than the reverse reaction.

Like esterification, transesterification needs a catalyst, in this case normally an alkaline catalyst such as Potassium or Sodium Hydroxide (also known as lye).

Both the catalyst and the excess methanol are left over after the reaction mixed with the products. In the second section of this manual we will explore the practicalities of separating them back out.

free fatty acids

When we make biodiesel from 'fresh' vegetable oil, we can assume that the oil is in the form of triglycerides only. Waste vegetable oil is a different matter however. When the oil is repeatedly heated, the triglyceride bonds are broken and some of the fatty acid chains break off. They are then known as free fatty acids (FFAs). Free fatty acids are a problem because they make the fuel acidic, which will damage the engine. There are various ways to deal with them. The most thorough method is to esterify the FFAs using the esterification reaction we saw earlier, giving methyl esters. This uses a concentrated acid catalyst. We can then transesterify using a basic (alkaline) catalyst as for fresh vegetable oil. However for everyday simple biodiesel making this two-stage acid/base reaction is a little complicated. Instead, we deal with the FFAs by turning them to soap.

$$R-\overset{\overset{\textstyle O}{\|}}{C}-OH \ + \ K-OH \ \longrightarrow \ R-\overset{\overset{\textstyle O}{\|}}{C}-O-K \ + \ \overset{\textstyle H}{\underset{\textstyle H}{\overset{\textstyle |}{\underset{\textstyle |}{O}}}}$$

fatty acid   potassium hydroxide   soap   water

The soaps end up mixed with the glycerol by-product. This reaction has the advantage that we can do it as part of our transesterification by adding more of the lye we are already using as a catalyst. We need to measure how much lye to add to turn the FFAs to soap and add that to the amount we need to catalyse the transesterification. To do this we **titrate** the oil.

We will learn how to titrate in the second part of this book.

making biodiesel

health and safety

Before commencing any practical work, it is of the utmost importance to understand the risks involved with making biodiesel. Safety can only be achieved when the biodiesel maker has been fully briefed regarding hazards and the safe working practices that ensure risks are minimised. Here follows a risk analysis that provides a safety model for small-scale biodiesel making.

risk analysis

Biodiesel production involves using:

1 dangerous chemicals
2 slippery oils
3 heavy containers
4 the application of heat
5 electricity

Each of the above aspects will be dealt with in more detail.

dangerous chemicals

Please see health and safety datasheets in appendix 3.

The following chemicals will be used:

1 methanol (CH_3OH) – technical grade 98-99%. This is a clear, colourless, odourless liquid (see data sheet for more details). It is highly flammable and has the added problem that the flame is invisible. You can become intoxicated on the fumes and also absorb it through your skin. Methanol kills receptor nerves, which means that you won't necessarily feel it burning your skin. Always work in a well-ventilated area with gloves, long-sleeved top and eye protection. Ensure that no hot surfaces are present as a splash could auto-ignite. Do not lean over large containers as the fumes are odourless and intoxication could result. Do not smoke in the presence of methanol. Make sure that you have a source of running water nearby. Wash off any methanol that comes into contact with skin or eyes with copious amounts of water, and seek medical advice. Note that conventional vapour masks are of no use when dealing with methanol however serious

they look. Methanol handling should wherever possible use sealed systems.

2   potassium hydroxide (KOH) – 100% (flake). This is a white flaked solid (at room temperature) that is violently caustic (see attached data sheet). It will cause severe burns if it contacts your skin. The tiniest speck in your eye will cause terrible damage and could easily result in blindness. The best case scenario is a three-month heal period. Always wear gloves (chemically-resistant nitrile rubber) and eye protection (preferably full face protection) when handling this chemical. Also, wear an apron and a sleeved shirt. If handling a larger quantity (if transferring from a sack to smaller containers for example) then also wear a mask. It is preferable to sodium hydroxide because the flakes are heavier and less likely to get airborne. It is also slightly less toxic than sodium hydroxide. Potassium hydroxide can be neutralized with vinegar (acetic acid). Keep a squeezy bottle (maybe a washing-up liquid bottle) of vinegar handy at all times. If any specks fall outside of their intended vessel then give them a squirt of vinegar. Immerse all implements in vinegar after they've been in contact with either the potassium or sodium hydroxide (or methoxide as described later).

3   sodium hydroxide (NaOH) – 100% (granules). Our process descriptions are generally oriented towards the use of potassium hydroxide rather than sodium hydroxide. Risks are as for potassium hydroxide but the size of the grains make it more hazardous.

4   sulphuric acid (H_2SO_4) – 98%. Our process does not use sulphuric acid, but this information is included in case you intend to use other recipes which require it. It is an orangey-yellow liquid that smells of sulphur. It is very acidic and will cause severe burns to skin and eyes. Wear an apron, nitrile rubber gloves and full face protection. Avoid inhaling fumes.

All of the above chemicals must be kept off the skin, and out of your eyes, and must not be inhaled – whether vapour or dust. It is important to read the data sheets.

These chemicals are best decanted into small containers in a lab situation. This minimises the risk of spillages that are difficult to manage. Spill management equipment must be at hand in sufficient quantities to contain the quantities that are present. This can be as simple as a bin of sawdust, a broom, and a scoop. Also, an eyewash station must be available with tops 'cracked' and ready for use, and in addition a source of running water nearby. We like to have a hose fitted with a trigger spray attachment constantly to hand. Take care not to splash liquids when stirring or mixing.

Store chemicals in a cool, dry, locked area. Use labels to denote the contents in containers. There should also be labels on the store room door to inform people of the potential hazards.

emergency first-aid procedures relating to biodiesel

- **ingestion:** rinse mouth with water and seek medical help
- **eyes**: rinse with water for 15 mins and seek medical help
- **skin**: wash with soap and water

slippery oils

The use of any oil presents a risk of slipping – either slipping over, or losing grip on things due to slippery hands. Making biodiesel can be a very messy business. Floors must be kept free of oil either by covering any spills with an absorbent material e.g. cardboard / sawdust, or with the use of oil absorbing granules. Glassware must be continually wiped (sometimes with solvent) to keep it free of oil. Handling glassware with oily gloves presents a very real risk of dropping containers and spilling dangerous chemicals. Use paper wipes to clean containers.

heavy containers

Care must be taken when lifting any heavy item. Lift with bent legs and a straight back. If an item is too heavy for an individual then use team-lifting techniques. Keep the load as close to the body as possible.

application of heat

During the biodiesel production process, heat is used to de-water and heat oil. This involves heating cooking oil to over 50°C. Great care must be taken to avoid spilling hot oil. Only use heating appliances that don't involve flames e.g. electric elements or hot plates or remote boilers transferring heat via a liquid. Use thermostats to control maximum temperatures where possible. Don't leave vessels of oil during heating – it's easy to become distracted and forget them. Install a smoke alarm in the laboratory area. Always have a source of cold water to quench burns. Don't overfill vessels with oil – it tends to spit and writhe on heating.

electricity

Use a power supply that is protected by an RCD (circuit breaker). Keep the work area clear of trailing leads that may make a trip hazard or knock things over. Be careful not to sever leads with heavy sharp objects.

other sensible precautions

1 A-well designed work area goes a long way towards minimising risks. Keep the heat and oily operations (dirty area) well away from the chemical (clean) area. Ensure that there is enough space, and stable work surfaces.

2 Keep a CO_2, foam (don't use foam on an electrical fire) or dry powder fire extinguisher handy at all times (near the exit). Be aware that the blast from a CO_2 extinguisher can knock over a burning vessel making matters worse – use it from a distance slowly approaching the fire. If you don't feel confident to tackle the fire then follow the procedure for evacuating the building and call the fire service.

3 Keep all rubbish in a metal lidded bin.

4 Be mindful at all times. Tell other people in your building what you are doing.

5 Dispose of rags soaked in biodiesel. Because biodiesel is made from plants, it can oxidize over time, which can produce heat – theoretically enough heat in a pile of rags to spontaneously combust and cause a fire.

making simple biodiesel

In this section we will go through the process of making biodiesel using a simple one-stage base-catalysed transesterification. This is the method used by many biodiesel 'homebrewers' throughout the world, with many variations possible in the detail. With care, biodiesel made by this method will work in any Diesel engine. It will fit the criteria for the lower duty rate in the UK though it is unlikely to meet all the stipulations of the European Biodiesel Standard EN14214. Later on in the manual we will look at some elaborations on this method that can improve the quality of the final product or allow you to use different feedstocks.

raw materials

feedstock

The most important raw material is of course the feedstock. This process will work with fresh oil, but we are presuming that you will be using waste oil (WVO). New oil costs around 46p per litre in UK (by tanker load), making it too expensive for biodiesel manufacture. WVO costs around 20-25p per litre from collectors who usually sell by the tonne, though larger collectors will not want to deal with small biodiesel projects. As regulations regarding use of WVO in animal feed are changing, the price of WVO from collectors will also change, and possibly their willingness to work with small biodieselers.

Most small-scale biodieselers will want to collect oil themselves. This is generally quite easy as catering outlets are often charged for disposal by collectors (about £1 a can; we have spoken with a caterer who was paying £1.50!) and are more than happy for you to take their oil away for free. Many of them will admit in private that they dispose of their waste oil by improper and illegal methods. Either way you are doing them a favour by collecting it. Bear in mind that larger outlets may be tied in to contracts with waste oil collectors who might not appreciate you on their turf. Most restaurants and fast food outlets store WVO in the 20-litre drum it came in originally. If you have a choice, try to get light oil that is very liquid: Chinese restaurants are a good source. Some caterers may use Palm oil that makes biodiesel that gels when it gets cold. Others may use animal fats. While it is perfectly possible to make good biodiesel with these feedstocks its better to start with easier, more liquid vegetable oils. You may also encounter very dark oil that might be the result of over-heating or overuse. Avoid this oil as it's very high in FFAs. You will also probably want to avoid these restaurants from a health point of view! You can expect to get 2 x 20 litre cans from the average fast food outlet per week.

Bear in mind that it's probably technically illegal to collect WVO for reasons we go into in the 'regulations' section of the manual. As you may well be collecting oil that could end up in landfill or rivers there's no reason to feel too guilty about it though. You need to prepare yourself with gloves and dirty clothes, as it can be a messy job. Try to make sure that no oil is spilt during transport and storage: it's surprising what an environmental hazard vegetable oil can be, especially if it enters watercourses.

alcohol

Biodiesel can be made with either methanol (methyl esters) or ethanol (ethyl esters). We use methanol for a number of reasons:

- the reaction is much easier to do with methanol
- methanol is cheaper
- there is more available research on methyl esters

In some ways it would be better to use ethanol from an environmental perspective. Even though methanol was originally produced from wood biomass it is now almost exclusively produced from fossil fuels, whereas it is possible to get ethanol from non-fossil sources. We're hoping that it will become possible to get biomethanol in the UK at some point so that the climate change impact of methyl esters is reduced still further. Making ethyl esters remains an advanced technique.

Methanol is available from a number of chemical suppliers. You may have trouble with some who are unsure about delivering to a non-industrial address. Occasionally it's possible to get methanol from other sources, such as drag racetracks or other biodiesel enthusiasts. Methanol is sold in model shops as model plane fuel but the cost is prohibitive in these tiny quantities. Generally you need to buy a 200l barrel to get a decent price: 40p per litre is a good price at the time of writing. 98% purity methanol or 'technical grade' is what you are after.

lye

The lye we use to catalyse the reaction and neutralise the FFAs can be sodium hydroxide (NaOH) or potassium hydroxide (KOH). NaOH is somewhat cheaper, compounded by the fact that less is needed for the reaction, but we use KOH for a number of reasons:

- KOH makes for an easier reaction
- KOH is slightly less dangerous
- not much is used, so the price difference is not so significant

NaOH and KOH are available from chemical supply houses. Ideally we want technical grade (98%). Supply companies are likely to be less worried about supplying lye than they are about methanol. Typical prices are around £8-9 per 20kg for sodium hydroxide (technical grade, delivered in a plastic sack); and £25 per 20kg for potassium hydroxide. 20kg will catalyse about 2000 litres of fuel so catalyst cost is around 1p/litre.

Only use fresh chemicals. Buy as much as you're going to use in the coming few months, and use them, don't store them – they're dangerous, and they go off. Sodium / potassium hydroxide is carbonated by CO_2 in the air, for example.

Lye can be very dangerous (see 'health and safety'), especially if it gets into the eyes. One grain in the eye is enough to cause terrible damage, and a minimum three-month healing period (if it heals at all). KOH tends to consist of bigger, more visible and less mobile flakes, and is therefore safer. You can also get both types of lye in pellet form.

NaOH is sometimes sold as drain cleaner but is unlikely to be very pure. We have made successful batches with own-label 'Caustic Soda' from B&Q though.

equipment

The equipment you need varies with the scale on which you make biodiesel. In order to describe the process we will suppose that you are making 'mini batches' based on 1 litre of WVO. For more information on scaling up the process see the 'small-scale reactor' section.

You can get the equipment from a number of sources including DIY shops, online tool catalogues, laboratory suppliers, jumble sales, your loft etc. In the following list we have suggested possible sources for some of the items in square brackets

safety equipment:

- eye-wash
- goggles
- chemical-resistant (nitrile) gloves
- vinegar in squeezy bottle [washing-up liquid bottle]

- hose connected to tap, with spray nozzle on other end
- sawdust to absorb spills [from sawmill or timber yard]
- paper towels or rags

titration kit:

- 20ml beaker
- isopropyl (rubbing) alcohol [lab supplies or pharmacy]
- de-ionized water [car parts or DIY shop]
- pH meter or phenolphthalein indicator (not universal indicator)
- calibrated pipette or syringe
- stirrer
- precise scales (to 0.1 g)

mixing and reaction:

- 250ml stoppered flask (glass) [lab supplies]
- hot plate (electric) [jumble sale]
- petri dish or similar
- 1500ml calibrated beaker
- strong jar around 2litre. [e.g. bulk mayo jar]
- old blender or drill with paint stirrer
- old saucepan
- thermometer

other useful stuff:

- pen and paper
- assorted containers and vessels
- sticky labels
- ladle

the process

Caution: first, make sure you have read the health and safety section, and that you are wearing goggles, mask and chemical-resistant gloves.

step 1: WVO pre-treatment

You want to start with about 2 litres of WVO for a 1-litre mini-batch, so we can pull off the best oil. The oil for your first batches should be liquid at room temperature. If you have some oil stored use oil from the top as the heavier fats move to the bottom over time.

Water in the oil will cause problems with the reaction. Too much water combined with the soap formed during the reaction could form an emulsion and your biodiesel could end up looking like chicken soup. For mini-batches we heat up the oil to over 100°C to boil off the water. Don't heat it too much though or you may char the oil, and create FFAs. Be careful: do this outside as the oil will spit and writhe.

When making full batches we use a different dewatering method: raise the temperature of the oil to 50-55°C and allow to cool slowly (you can insulate the vessel with an old blanket or loft insulation) and settle. Water won't settle during heating because of convection currents. Water will fall to the bottom of the vessel because it's heavier than the oil, and can be tapped off the bottom. This method uses much less energy than trying to drive it off by heating to over 100°C, and is also quite effective.

Pour the oil into a jar through a sieve to remove any large bits like chips, cigarette ends, or large visible chunks. You can also use a J-cloth as a filter. This does a good job of getting smaller particles out of old oil, though it can take some time to drip through.

step 2: titrate

In the chemistry section we saw that we use lye (in this case, KOH) for 2 purposes: as a catalyst for the transesterification and to neutralise the FFAs.

The amount of KOH we need to perform the transesterification is fixed: for every litre of oil, you need 9 grams of KOH (or 3.5 grams of NaOH). This amount has been worked out using the transesterification formula based on the average composition of vegetable oils.

If we were using fresh vegetable oil, we would just use 9g of KOH. But with WVO we need to add an amount to neutralize the FFAs. This amount will depend on the oil in question. To work out how much we need, we perform a titration.

For every batch of oil, you need only one titration, although you do need to mix the oil well first, to make sure that it is homogenized, and so has the same properties throughout. You do, however, need to do a titration for every new batch of oil, as the free fatty acid content of each batch will be different.

It's best to perform the titration in a warm area. Whenever you do a titration you will need a standard solution of KOH in distilled water. We will start by preparing this. It should last for a long time if kept stoppered.

making standard solution

1. weigh 1 gram of KOH on a petri dish on a scale
2. take a 1 litre stoppered container and fill with de-ionised water. You will probably have bought the de-ionised water in such a container
3. pour the KOH into the distilled water and swirl well to dissolve. It should dissolve fairly easily
4. label this bottle 'KOH standard solution'

titration

1. Add 10ml of isopropyl alcohol to a beaker.
2. Add a few drops of phenolphthalein and stir.
3. Use a calibrated pipette / eye dropper to add 0.5ml of the KOH standard solution into the oil / alcohol and stir.
4. Observe the colour of the mixture. You will notice a flare of red as you add the KOH solution, which will disappear as you stir.
5. Continue to add the KOH standard solution, 0.5ml at a time, until the red colour persists after stirring. This occurs when the isopropyl alcohol itself has been neutralised. This is called a 'blank titration'; we have now accounted for the acidity of the isopropyl alcohol.
6. With a calibrated pipette / eye dropper, add 1ml of the WVO to the isopropyl alcohol and stir. It's best to take 1ml of oil

from your sample after you've filtered and de-watered it. You need to dissolve the oil in the alcohol completely. You will notice that the red colour disappears again: the FFAs in the oil have made the mixture more acid. At this point you start keeping a record of the amount of KOH solution you are adding.

7. Continue to add the KOH standard solution, 0.5ml at a time, until the red colour persists after stirring. (this is when the free fatty acids have been neutralised). The exact colour will vary with the colour of the WVO: what you are looking for is a definite and persistent colour change, normally to a brick red or purple.

8. Note down the amount of KOH standard solution you have added since completing the blank titration; this figure will normally be no more than 3ml.

We used to use a pH meter to perform the titration. The technique is similar: you keep adding KOH standard solution and plotting pH readings on a graph to find the point at which the pH rapidly rises . We have moved to using phenolphthalein on our courses and in the lab because cheap pH meters are often inaccurate and all pH meters need careful cleaning and recalibration. However you titrate, remember that you are looking for a sudden change in pH as the FFAs are neutralised. This neutralising point is not at pH7 because we are neutralising a weak acid with a strong base which forces the pH upwards ('the weak acid is incompletely ionised in solution' for all you chemists!). Don't worry too much about precise pH: it's the sudden change we are looking for. Some biodiesel recipes leave out the blank titration stage, but when you have got the hang of it, it only adds minimal complication for a more precise result.

calculations
The number of ml of KOH standard solution it takes to neutralise the FFAs in 1ml of WVO equates to the number of grams of KOH it takes to neutralise the FFAs in 1 litre of the same WVO.

x ml of KOH standard solution for **1 ml** of WVO
x litres of standard solution for **1 litre** of WVO
x litres of standard solution contains **x grams** of KOH

This is rather handy! To find the total amount of KOH we need we just add the amount needed as catalyst for fresh oil (9g) to the result of the titration. So if the titration endpoint was reached after 2.5 ml of KOH standard solution was dripped into the oil/alcohol mixture, the total amount of KOH needed would be 9+2.5 = 11.5 gram per litre of WVO

That baseline amount of 9g is the amount of KOH needed to catalyse the reaction . This will actually be different for different feedstocks but a figure of 9g gives good results in our experience across all the types of WVO we have worked with. If you are using NaOH (Sodium Hydroxide) the baseline amount is normally given as 3.5g. Of course if you use NaOH you should also titrate with NaOH standard solution rather than KOH.

step 3: mix methoxide

For every 1 litre of oil, you will need 200ml of methanol. This needs to be mixed with the amount of KOH we calculated from the titration.

Put 200ml of methanol in the stoppered flask. Weigh the correct amount of KOH and add to the methanol. Swirl or stir the flask until all the KOH is dissolved in the methanol. You will notice that as the KOH dissolves it will release heat and the flask will become warm.

Take care with this step as it involves the two most dangerous chemicals we use in simple biodiesel making. Mixed together they're no nicer!

Biodieselers often call this mixture methoxide. Remember that it is not a chemical compound but a mixture: there are no bonds between the KOH and the methanol.

You might be wondering about the 200ml figure. Like the baseline amounts given for catalyst, this 20% methanol figure is subject to some debate. In large-scale biodiesel manufacture, 25% is often used. On the other hand, in making crude biodiesel, quantities as low as 12% have been used. In general the more you use up to 25%, the more complete the transesterification reaction is, but the more excess methanol you have to recover once the reaction is over.

step 4: reaction

Add the methoxide you have prepared to the oil. Although it will work at a range of temperatures, the reaction works best if the oil is at about 45°C. Too much higher and the methanol (which boils at 65°C) starts to evaporate off. As soon as the reactants are mixed, you need to start stirring the mixture with the blender or other stirring device. A violent stir in the first few minutes of the reaction ensures best conversion to biodiesel.

You need to be careful to avoid breathing in methanol vapours. Also if you are using an electric motor to stir the reaction, be aware that a spark from the motor could ignite methanol vapours. Never use a drill or similar motor on anything bigger than a mini-batch.

On a mini-batch, 10-20 minutes of stirring will easily suffice. On larger batches it may be hard to get efficient stirring and some biodieselers stir for 2 hours or more. When oil and methoxide are first mixed, the mixture is a milky colour; when it's ready, it's dark – look for the colour change as an indication that the reaction is complete.

step 5: settle

Once the reaction is complete you need to leave the mixture to stand overnight. The next day, a thick brown liquid will have settled at the bottom. This is glycerine, and the lighter-coloured liquid above it is biodiesel. Biodieselers tend to use the term 'glycerine' quite loosely: in this layer there will be glycerol, most of the surplus methanol, most of the catalyst, and most of the soaps formed when the FFAs were neutralised. You can expect to have around 120-180ml of this 'glycerine' layer, which is more accurately called the 'by-product phase'. Of this, perhaps 80ml will be glycerol, up to 80ml could be methanol, the rest is mostly soap.

step 6: separate

The biodiesel and the glycerine are easy to separate. In a jar you can pour most of the biodiesel off the top as the glycerine is much more viscous. In a reactor you would draw the glycerine off the bottom.

Congratulations: you have made biodiesel! Or have you? There are a few simple tests you can perform to see whether your reaction has been successful

step 7: test

The simplest test is the visual test. You are expecting to see a very clear distinction between biodiesel and glycerine. If that is present then you have most likely succeeded to some degree. If you have a very small amount of glycerine (i.e. less than 100ml) then it is likely that your reaction did not proceed to completion. The likely cause is using too little catalyst. You might want to check the titration again, or reprocess your product to see if more glycerine will come out. If there are large amounts of visible soaps between the two layers you may have over-catalysed or failed to de-water the oil correctly. Over-catalysing will occasionally lead to the formation of a gel or emulsion.

You should have about the same volume of biodiesel as you had vegetable oil in the beginning, or a bit less if it was high in FFAs, as these will have turned to soap.

You can do simple tests for density and viscosity. Using a hydrometer with a scale from 800-1000 you can check the density of your fuel. It should be between 880 and 900 (that is, your biodiesel should weigh between 880 and 900g per litre). You can also check this with a set of scales.

It is not easy to get an absolute measure of viscosity but you can get a comparative measure. Pour a measured amount of biodiesel into a tin with a hole in it and time how long it takes the tin to empty. Do the same for a sample of the original WVO and for a sample of petrodiesel if you have some. The biodiesel should go through the hole much faster than the original WVO, showing that you have succeeded in reducing the viscosity. The measured times for the biodiesel and the petrodiesel should be similar, though don't worry if your biodiesel is a little more viscous.

step 8: wash the fuel

At this stage you have succeeded in making a small amount of basic biodiesel. You could decide to scale up production using the same method, and in fact this is what many homebrewers across the world have done, some of them with reported success. However the fuel we have made is still rather crude. Not all engines will run happily on it as it stands, or if they do, they may not continue running for all that long! Although most of the impurities are in the glycerine, there are still small amounts of lye, soap, methanol and water in the biodiesel. These will cause trouble in your fuel system and engine. The easiest way to remove them is by washing the fuel with water. You can try this with a mini-batch. Add water to a jar containing your biodiesel. Add the water carefully, dribbling it down the side of the glass. You will notice that the biodiesel floats on top of the water (remember that it is less dense than water: the density is 0.88 to 0.9 percent that of water). Keep adding until you have about 300ml of water to your litre of biodiesel. Now stir the mixture quite gently, but enough so that the two layers are mixed together. You will notice that the water becomes cloudy. As you mix, the lye, soaps and methanol are becoming dissolved in the water. This is because the water is better at dissolving the impurities than the biodiesel is: it is a more polar solvent.

You need to mix carefully because of the presence of soap. Soap is an emulsifier: if you mix too violently the biodiesel and the water will become bound together in an emulsion: the dreaded 'chicken soup'. As you wash, more of the soap becomes dissolved in the water.

Carefully remove a sample of the biodiesel from the top of the jar using a ladle or similar. Add this to another jar and repeat the washing process. This time you will find that you can stir more violently without forming an emulsion, as much of the soap will have been removed. One of the less obvious benefits of water washing is that as the soap is removed from the biodiesel, the biodiesel becomes less able to bind water molecules, so in effect we are using water to dry our fuel! Water washing also removes the small amount of free methanol in the fuel. This stops the reaction which would otherwise continue very slowly, dropping out small amounts of glycerine which can clog the fuel filter of your vehicle.

The initial results of water washing are often rather disappointing. The lovely transparent golden liquid you found after your morning's settling has been transformed into a straw-yellow and somewhat cloudy liquid. Don't panic. The haziness is due to a small amount of water in the fuel which will quickly settle out. Just take another look at that wash water and you can no longer doubt that washing was a good idea. Even better - test the pH. It will normally be basic (pH >7). We want our biodiesel to be as near neutral (pH 7) as possible.

step 9: filter and de-water the fuel

It's always necessary to give the fuel a final filtration before using it in a vehicle. Although the fuel system incorporates a fuel filter which removes particles and traps water, it's best not to overburden it. The simplest way to filter a mini-batch is to pass it through a J-cloth or similar tightly-woven cloth. This filters to about 10 microns. Even If you already did this with the WVO at the beginning of the process, it is still worth doing again as there are plenty of opportunities for unwanted bits to get into the fuel during the process.

Biodiesel, like petrodiesel, is hygroscopic, meaning that it absorbs water from the atmosphere. Normal biodiesel has 1200-1500 ppm (parts per million) of water. Any water above this level will damage your engine if it gets past the fuel filter. If your fuel is well washed, it will not hold on to much water. However if it remains cloudy after a day or so of settling, it is advisable to de-water. You can do this using the same technique as for WVO: heat to 50-55°C and allow the water to settle out.

step 10: deal with the by-products

Wash water contains soaps, lye, and a small amount of methanol. In mini-batch size quantities it should be ok to put this down the drain. You can think of it as bath water into which you accidentally spilled some of your bathtime whisky.

The glycerine will contain soaps, glycerol, lye and quite a lot of methanol. The 150-odd ml you get from your mini-batch can be left somewhere well-ventilated to evaporate the methanol, after which you can use it to make soap or compost it.

Of course in larger quantities the by-products need to be dealt with more comprehensively. We will discuss this in the next section.

practical biodiesel making: a 100-litre batch reactor

In order to look in more detail at the issues involved in practical biodiesel production we will take the example of a small-scale batch system making 100-litre batches. This sort of system might be appropriate for an individual, family or small community making biodiesel for their own use.

This reactor has been designed to use readily available materials to make a small-scale batch processor (see photo below).

materials

Most of the processing takes place in the two tanks. It is generally said that stainless steel is the best material for biodiesel plant but it is expensive and hard to work with.

Some commercially available systems are made from plastic, normally HDPE (High Density Polyethylene). This is a cheaper option, and has the advantage that off-the-shelf tanks are available with conical bottoms. This allows for accurate and easy separation of different layers or 'phases'.

However we have found that standard oil drums made from mild steel are suitable for a simple processor. They have the advantage of being readily available, very cheap, and easy to work with. They also stand up well to high temperatures, and give a degree of containment in case of fire. It is possible to weld conical bases onto them but in this design we have simply dented the bases, which works well enough.

Mild steel can be coated with chemical-resistant paint: some steel drums are already coated in a suitable paint. Oil drums are roughly 205 litres (45 UK gallons or 55 US gallons). Some biodieselers have used copper central heating cylinders as main reactors, because they already have useful fittings and a conical bottom when turned upside down.

Pipework is made from 22mm copper pipe with standard bronze compression fittings. In theory it might be better to use chemical resistant plastic hose but the fittings are expensive and hard to find. Bronze has very good chemical resistance, copper less so, but the liquids do not stay in the pipes for long enough to cause a problem. We use compression fittings not solder as the solder is not chemical-resistant. Joints are sealed with high-temperature silicon sealant or PTFE tape.

The heart of the system is a multifunction pump. It is used to transfer treated oil to the reactor, to mix the reaction, and to evacuate the finished product. This inexpensive water pump has a bronze impeller and a totally enclosed fan-cooled motor, which contributes to safety (no spark risk). A system of taps allows liquids to be diverted through particular pathways.

Heating is provided through standard domestic immersion heaters. These are attached to 2¼ inch flanges that can be brazed on to the tank. In our case we have used mechanical flanges, which are held on with a screw thread.

The electrical appliances are wired with heat-resistant flex into a control centre with illuminated switches. The unit runs off a standard 13amp socket and is equipped with a RCD plug. All metal parts are connected and earthed.

oil preparation

Good oil preparation is essential for making decent biodiesel. In this processor we have incorporated an oil preparation vessel into the design. Waste oil is poured through a sieve into the top of the oil preparation vessel. The immersion heater in the side of the tank is then switched on to bring the oil up to 50-55°C. With the lid on and an insulation jacket, this takes around an hour. It is also possible to heat via a heat exchanger if you have a readily-available source of heat. Some biodieselers run a generator on their own fuel and use the waste heat for this.

We then allow the oil to stand by opening the valve at the bottom of the vessel to draw off the water along with many of the solid impurities which made it through the sieve. We continue to draw off the nastiest oil from the bottom until the top of the oil reaches a pre-measured mark. At this point we know we have 100 litres of oil. This oil preparation vessel has an out-pipe positioned some way up the side of the vessel. The oil that leaves the tank is not drawn from the bottom but from this out-pipe. This means that residual solids at the bottom of the tank will not be drawn out with the oil (and gets round the issues associated with not having a conical base). Where oil passes out of the tank it goes through a simple in-line filter. This is a standard and inexpensive plumbing part, which has a washable stainless steel 400-micron filter. The fuel will be filtered again at the end of the process, but this pre-filtering avoids damage to the pump and unnecessary clogging of the final filter.

methoxide preparation

While the oil is heating up we can prepare our methoxide. This is one of the most hazardous aspects of biodiesel production, as it

involves manipulating significant quantities of hazardous chemicals. Therefore it is essential to do as much of the mixing process as possible in a sealed environment. In this processor the methanol is first pumped into the methoxide vessel from the methanol storage drum. We are using a methanol-resistant hand pump made from Ryton. This screws into the methanol drum. Note the gloves, glasses and overalls. The operator is also wearing a mask, which is no protection from methanol vapour but will be useful for the next part of the process. Using the funnel that you can see in the bottom left of the picture, the weighed lye is then added slowly to the methanol. This dissolves slowly over the time it takes the WVO to warm up.

There are a number of different strategies for methoxide preparation. Because we use KOH the lye dissolves fairly easily in the methanol. With NaOH it takes longer to dissolve. In both cases it is helpful if you can give the mixture a stir. Some processors incorporate mixing paddles operated by electric motors into their design. This ensures the methoxide is well mixed but adds to the cost and the possibility of sparks near methanol. One of the best approaches is to mix the methoxide in a separate 20l can that can be closed and then gently swirled by hand. It is also possible to make a large quantity of methoxide ahead of time when using KOH. With NaOH the methoxide goes off much faster. Some people keep two barrels, one containing methanol and the other methoxide prepared to a standard concentration. By mixing the two liquids they can create methoxide to match any titration.

the reaction

The reaction vessel is where the transesterification takes place. It needs to be well sealed to prevent the escape of methanol during the reaction. It must incorporate some way of stirring the WVO and methoxide together. In this design we use a pump to mix the reaction. Other designs often use large paddle mixers to stir. However using a pump gives a good mix and avoids the need for more electric motors.

In the past we have prepared the methoxide in the reaction vessel, used the pump to mix it, and then added the oil. In this system the methoxide is sucked into to the flow of oil. This improves the mixing as the methoxide gets churned up in the pump immediately after entering the oil stream. Some small-scale set-ups use a venturi fitting to suck up the methoxide. A venturi is a fitting which uses the flow of liquid to pull a vacuum. The advantage of this is that the methoxide container does not have to be above the point of mixing: it can be safely on the floor. Unfortunately venturi are hard to make and expensive to buy.

It's best to insulate the reactor while the reaction takes place so the residual heat from the WVO preparation is not lost. You can use bubblewrap or loft insulation, though be careful as the plastics involved are often not up to the temperatures. It is also possible to heat the reactants with an immersion heater or heat exchanger, though in this system this is not necessary or indeed possible as using the pump and heater at the same time would use more power than an ordinary 13 amp electric outlet can supply.

methanol recovery

Having reached this stage we normally leave the fuel to settle so we can remove the glycerine. However this reactor incorporates a condenser to recover the excess methanol at this point in the process.

Many homebrewers do not do methanol recovery. There are several reasons for this:

- it has a high calorific value, and is therefore a good fuel itself – improving range and power
- it thins the biodiesel, improving cold starting, and reducing the risk of coking injector heads
- it helps preserve the fuel in long-term storage, and stops bacterial growth

We think recovering methanol is important because it will otherwise end up in the glycerine and the wash-water as well as the fuel. We spilt wash water with methanol in on some grass 2 years ago and nothing has grown there since! There are other good reasons to recover methanol:

- so that the lower rate of duty is payable on your fuel
- to use it again; financially this is a good idea, as the methanol recovered will be more valuable than the energy used to remove it
- methanol can damage the sensors in new cars

In a 100-litre batch using 20% methanol we have about 8 litres of methanol to recover.

To operate the still, the reactor is left covered and the immersion heater in the tank is switched on. This brings the tank up to 80 degrees, which is above the boiling point of methanol. The tap leading from the top of the reactor to the condenser is opened. Methanol vapour enters the condenser, where it cools rapidly, turning the vapour back to methanol liquid. This drips down into the methoxide vessel. As the concentration of methanol in the mixture drops, the temperature has to be raised in the reaction vessel.

Recovering methanol at this stage has several drawbacks. You have to make sure the reactor is very well sealed or methanol vapour will escape. It takes a lot of energy to heat that much fuel and it's hard to control the temperature precisely enough. There is also a risk as you remove the excess methanol that the reaction will start to go backwards, turning your precious fuel into fatty acids or triglycerides.

A better approach for the homebrewer may be to recover methanol from the glycerine only, after settling (see later section titled 'What to do with the glycerine). This works because the majority of the methanol is in the glycerine. The small amount left in the fuel comes out in the wash water. There is a trade-off to be made here between fuel quality, environmental considerations, and effort. You could of course recover methanol from both the glycerine and the fuel separately but recovery from the fuel is an extra step involving energy and effort for little return. Another way of dealing with this problem is to use less excess methanol for the reaction. This means that the reaction will not proceed all the way to completion. Your product will include unreacted triglycerides and partly-reacted triglyceride in the form of mono- and di-glycerides. This is a lower quality fuel but will work well in many more fuel-tolerant vehicles. The advantage is that there is less excess methanol to worry about and the amount which ends up in the wash water may be trivial.

As you can probably tell, we do not feel that the definitive homebrew system for recovering methanol has been designed yet. Small scale biodiesel kits on the market tend to ignore the issue completely. We are continuing research on this; it's probably the biggest technical challenge in small-scale biodiesel production.

settling

Now the reaction is complete we can leave the products to settle out. Glycerine settles out best if the mixture is warm, so it's good to keep the reaction vessel well insulated. You could also decant the mixture into another vessel for settling, freeing up the reactor for another reaction. In order to separate the glycerine accurately from the fuel a conical bottomed vessel is best. In this system we have a crudely dished oil drum. When we draw off the glycerine we inevitably draw off some fuel with it. This can be recovered from the glycerine when we come to deal with that later. In large biodiesel plants the settling process is achieved with a centrifugal separator that spins the heavier glycerine to the outside of a chamber and removes it. Often the process involves many separator stages. As these are very expensive bits of kit we've seen smaller scale plants that attempt to use separators designed for cream separation and even a fuel/water separator from a submarine! These experiments have not been a success so we are content to use gravity and time.

water washing and dewatering

In order to get a good water wash we need to maximise contact between the water and the fuel without excessively violent mixing. There are two main techniques used in biodiesel plants. The first is

called mist washing. This involves spraying a very fine mist of water on top of the fuel . The tiny droplets dissolve the impurities in the fuel as they sink down through it. The water ends up as a layer underneath the fuel which can be drawn off the bottom of the vessel.

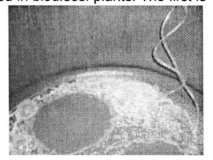

In this reactor we use bubble washing. First we add about 30 litres of water dribbled from a hose down the side of the vessel. Then we drop an aquarium air stone into the bottom of the vessel attached to a small air pump. The bubbles form in the water layer with a thin film of water around them. They then rise through the fuel, the thin film of water dissolving impurities in the biodiesel. At the top of the liquid the bubbles burst and the water travels back down the same as with the mist wash. The advantage of the bubble wash is that it can be left overnight, giving a very thorough wash. The air pump uses only a few watts of power.

Air pumps and air stones are easily available from pet shops. Some air stones dissolve in biodiesel: the ceramic white stones are more expensive but worth it. Some biodieselers make their own bubblers from grinding discs or perforated copper pipe.

Older recipes for making biodiesel will tell you to avoid bubble washing as it agitates the mixture too much and causes emulsions. They may also recommend the use of acid in the wash water to neutralise soaps. These precautions are generally only necessary if your original WVO was very high in FFAs and you have tried to neutralise these with a large amount of lye. In these cases it is better to use an acid/base method to esterify the FFAs. This is covered later in this section.

Some set-ups have a separate washing vessel. This is useful as it frees up the reactor. In this system we wash in the reactor for simplicity. The reactor vessel already has an immersion heater so we can use this to dewater the biodiesel if necessary. Some systems use zeolite to de-water. This is an inert substance that absorbs water, a bit like cat litter. You can put a bag of zeolite in the fuel to dewater it. The zeolite needs to be dried to regenerate it. This can be done in the sun or in an oven.

Another de-watering method is to bubble air through the fuel (bubble-drying, not to be confused with bubble-washing, but potentially using the same equipment). This may cause oxidation of the fuel however. On the biodiesel Internet forums the debates rage backwards and forwards and there is no 'definitive' technique for drying.

fuel dispensing

As the finished fuel is still in the reactor we can use the pump to dispense it into storage vessels or directly into the vehicle. On the way out it passes through a 5 micron filter. In this case we are using an ordinary water filter. You can also use vehicle fuel filters. It helps to use a pump at this stage as 5 microns is quite fine and gravity filtering can take a very long time.

variations on the basic method

This reactor is designed to use the single-stage base-catalysed transesterification method we have been describing in this manual. There are a number of variations on this method which may yield better fuel quality with certain feedstocks.

2-stage base/base method

In the chemistry section we mentioned that one way of pushing the reaction to completion was to retract the products, but that this generally required equipment such as separators. This is why we rely on excess methanol. One way to get round this is to split the reaction into two stages. The methoxide is prepared as normal and split into 2 equal parts. We perform a reaction as normal with the first half of the methoxide. We then settle out the glycerine overnight as usual .Then we react the product with the second half of the methoxide. This will pull out more glycerine. This method should result in more complete transesterification. Potentially it also allows us to reduce the amount of excess methanol. Variations on this technique exist with different proportions of methoxide used for first and second stages, and some people even do three-stage reactions like this

2-stage acid/base method

This method is useful for dealing with feedstock which is high in FFAs. The first stage involves treating the WVO to esterify the FFAs. To do this we need methanol and an acid catalyst such as concentrated sulphuric acid (nitric acid must not be used in biodiesel chemistry as it may create nitro-glycerine!). The FFAs are esterified and turn to methyl esters (biodiesel). The second stage is the normal base-catalysed reaction but the titration should show that the oil is now very low in FFAs. Less lye will be needed.

This reaction is much better for high FFA feedstocks. Instead of turning the FFAs to soaps which can cause problems with the washing stage, we turn them into biodiesel! The downside is that it involves handling concentrated acid and that it complicates the procedure somewhat. It has been touted as an easy or 'foolproof' method but it is really a more advanced technique. You need to master the standard method before trying this one.

There are also variations on this method. Some people reverse it, using base to turn all the WVO to FFAs and then esterifying the whole lot with an acid catalyst. This is useful for feedstocks like tallow or trap grease.

what to do with the glycerine

When you make larger amounts of biodiesel you will inevitably end up with quite a large store of glycerine. Each time you make 100 litres of fuel you can expect to get somewhere between 8 and 15 litres of crude glycerine. This glycerine contains glycerol (about 8 litres) and excess lye (the catalyst is always left over after the reaction). If you did a single-stage reaction with WVO, the crude glycerine will contain soaps formed by the neutralisation of the FFAs. If you have not done methanol recovery yet most of the methanol will be in the glycerine.

You would normally start by recovering methanol from the glycerine. This takes less energy than recovering it from the whole reaction mixture as there is less to heat. You will have to be careful as immersion heaters can easily burn out in glycerine: the heat does not spread easily. Most homebrewers who do this build a small still using water bath heating or a heat exchanger.

The equipment you need to recover methanol from glycerine is very similar to that used to distil alcohol. In fact it has long been the custom of moonshine-makers to throw the first cupful from the still onto the fire as it contains poisonous methanol. You can find a lot of relevant advice on websites dedicated to the art of backwoods distillation... bearing in mind that it is illegal in the UK to produce your own spirits for drinking purposes.

The basic principle is to bring about uniform heating of the glycerine inside a well-sealed vessel with a pipe coming out of the top. Methanol evaporates at 65C, but in practice you will need to get the glycerine to around 80C before significant quantities of vapour form. A small amount of pressure will build up, forcing the methnanol vapour out of the pipe in the top of the vessel.

The pipe leads to a condenser, which cools the vapour rapidly, turning it back into liquid methanol which drips into a collection vessel. The condenser can be as simple as a coil of flexible copper pipe in the air. If you are doing this on anything larger than a mini-batch basis you will probably need to cool the condenser coil with water or possibly with a large cooling fan. Some people use a pipe-within-pipe arrangement as a condenser.

Methanol recovery can be a time-consuming process. Some homebrewers accelerate it by using a vacuum pump to pull the vapours through the condenser. At all times it is worth bearing in mind that you are working with methanol vapour which is hazardous to health and extremely flammable. Any equipment you construct needs to be properly sealed and completely free of the possibility of sparks that could ignite the vapour.

It's easy to see why the commercially available small biodiesel kits invariably ignore the challenge of methanol recovery and just leave it in the glycerine. We are still experimenting with different setups in order to find a cheap and safe design. The best advice we can give is to do your own research, and to seek advice from others before implementing any design. Internet forums are probably the best medium for this.

Glycerol is very valuable in its pure form and large biodiesel plants always include equipment to purify it. Unfortunately this involves distillation techniques which are very complex and require expensive equipment. For the small-scale producer this is impractical. To sell crude glycerine you need to produce many tonnes. Therefore we have to find other uses for glycerine.

purification

Although we can't distil it, it is possible to purify the glycerine to the point where a soap-maker could use it. We can do this by adding phosphoric acid to the crude glycerine. This converts the soaps back to FFAs which separate out. The lye also precipitates to the bottom as a powder. The FFAs can be collected and esterified to make biodiesel. The remaining glycerine is about 95% glycerol.

burning

Glycerine burns well but must be combusted at a high temperature or it releases acrolein, which is a very hazardous gas. It is also possible to etherify the glycerine to make a fuel.

composting

Crude glycerine will compost well and the lye is a fertiliser. It will need mixing with cardboard or similar to avoid it turning into a slimy mass.

soap making

This is perhaps the most common use for crude glycerine on a small scale. You can make a simple liquid soap from crude glycerine: heat it to reduce viscosity, filter it, add about 5% essential oils to make it smell nice, allow to cool and stand in an open-top vessel for at least a week.

To make hard soap you can use materials you already have: KOH and water. Gloves and goggles on as always when handling lye.

1. Calculate the amount of catalyst you need to add to the glycerine: you need to use around 100g KOH for every litre of glycerine. Measure how much glycerine you have (in litres) at the end of your reaction, and multiply by 100 to see how many grams of catalyst you need. Take away the original amount of catalyst

that you used in the reaction (which is already in the glycerine) and this is how much KOH you need to add (in grams).

2. Calculate the volume of water you need to dissolve the catalyst: for every litre of glycerine you need 340ml of water. So multiply the number of litres by 0.34 to find the total amount of water you need in litres.

3. Heat the water to 32°C.

4. Slowly add the amount of catalyst you calculated in 1.

5. Stir until all the catalyst is dissolved

6. In a large pot, heat the glycerine to 45°C.

7. Add the catalyst solution slowly.

8. Continue to stir slowly and keep the mixture at 45°C for 10 minutes.

9. Add around 5% essential oils to make the soap smell nice.

10. Pour the mixture into a container lined with a damp cloth or tea-towel.

11. Put a lid or a piece of board on top and wrap in a blanket.

12. Leave it for 24 hours.

13. Pull the block out of the container using the cloth.

14. Cut into smaller blocks the right size for bars of soap.

15. Let the soap sit for at least a month before use.

storage

Biodiesel is much safer to store than mineral diesel. It is non-toxic, biodegradable, and has a higher flashpoint. Like petrodiesel, algae can grow in biodiesel causing slime to gunk up fuel lines and filters. Storage containers should be clean, dry and dark to avoid algae formation. Also like petrodiesel, biodiesel is hygroscopic: it will absorb water over time. Avoid copper, zinc or tin for long-term storage of biodiesel: mild steel, aluminium, stainless steel or polyethylene are good.

biodiesel standards

At the moment there is no need for small-scale biodieselers to meet any particular standard with the biodiesel they produce. In the UK there is a strong incentive to reach the Customs and Excise definition of biodiesel. This is discussed further in the 'regulations' section: the relevant stipulation is that in order to qualify for the reduced duty rate, fuel must have an ester content of not less than 96.5% by weight. Since vegetable oil and biodiesel are both esters, this is achievable even if the reaction does not go to completion. It does mean that methanol must be removed from the fuel though, as we need that extra 3.5 percent to encompass water, free fatty acids and mono- and di-glycerides.

It is possible that in the future this definition may be tightened. The European Standard for biodiesel (EN14214) might become the yardstick for the lower rate of duty. With the methods outlined in this manual it is perfectly feasible for the homebrewer to make fuel that meets EN14214. It does require care however and would probably use one of the more advanced two-stage techniques.

You can have one, all, or a selection of these tests performed by a lab. ASG Analytik in Germany specialises in biodiesel testing. Simple tests can give us a good idea how near we are to reaching the standards.

EN14214

Until recently, different countries had different standards regarding biodiesel, but in Europe, most countries were working towards the German standard (DIN51606). The UK has now adopted the European specification for biodiesel (EN14214). If vehicle warranties allow the use of biodiesel they usually specify that it must be made to EN14214 or DIN51606, which is very similar.

Criteria	Derv (EN590)	Biodiesel (DIN51606)	Biodiesel (EN14214)
Density @ 15°C (g/cm³)	0.82-0.86	0.875-0.9	0.86-0.9
Viscosity @ 40°C (mm²/s)	2.0-4.5	3.5-5.0	3.5-5.0
Flashpoint(°C)	>55	>110	>101
Sulphur (% mass)	<0.20	<0.01	<0.01
Sulphated Ash (% mass)	<0.01	<0.03	<0.02
Water (mg/kg)	<200	<300	<500
Carbon Residue (% weight)	<0.30	<0.03	<0.03
Total Contamination (mg/kg)	-	<20	<24
Copper Corrosion 3h/50°C	Class 1	Class 1	Class 1
Cetane Number	>45	>49	>51
Methanol (% mass)	-	<0.3	<0.2
Ester Content (% mass)	-	>96.5	>96.5
Monoglycerides (% mass)	-	<0.8	<0.8
Diglycerides (% mass)	-	<0.4	<0.2
Triglycerides (% mass)	-	<0.4	<0.4
Free Glycerol (% mass)	-	<0.02	<0.02
Total Glycerol (% mass)	-	<0.25	<0.25
Iodine Number	-	<115	<120
Phosphor (mg/kg)	-	<10	<10
Alkaline Metals Na. K (mg/kg)	-	<5	<5

explanations

- density @ 15°C: 0.86 – 0.9g/cm³ (880-900g/litre). Also known as specific gravity. You can check this with a hydrometer or scales as detailed above. You can easily make homebrew to the European spec. as regards specific gravity.

- viscosity @ 40°C: 3.5-5 mm³/sec. Viscosity is hard to measure. We can measure relative viscosity easily but this measurement is of kinematic viscosity. It's a lab test but the specs are easy to meet.

- flash point: >101°C. This is the temperature at which vapours above the liquid will ignite if there is a source of ignition. Petrol has a flash point of between 20 and 30°C and is therefore dangerous; the flash point of methanol is around 10°C – even more so. In reality, the actual flash point of biodiesel is well above the specification, at around 160°C

- sulphur content: 0.01%. This is a crucial one, as it is one of the criteria that you will have to meet to pay the lower rate of duty. It's not a problem for biodiesel or SVO – virtually no sulphur content. The same applies for sulphated ash.

- water: <500mg/kg (i.e. <0.5%) Water in diesel can cause cylinder pitting. Research has shown that petro-diesel bought from pumps occasionally contains as much as 1% water. The spec demands 0.5%, so water content is an issue – they would like you to remove as much water as possible.

- carbon residue: 0.03% of mass. Carbon that remains when the biodiesel is burnt

- cetane no: >51. Relates to how well the fuel ignites. This is no problem for reasonable quality biodiesel which has a higher cetane number than petrodiesel.

- methanol: 0.2% of mass. Comprehensive methanol recovery is essential to meet this standard.

- ester content: 96.5%. This is the standard that UK Customs and Excise got their figure from. Here 'ester' means 'methyl ester'. This is one of the hardest standards to reach as it requires the reaction to go almost all the way to completion.

- mono-, di- and triglycerides: <1.4% of mass in total. This standard essentially clarifies the previous standard by saying that the unreacted and partly reacted oil has to be very low.

- CFPP: (no figure for spec yet). The cold filter plug point is the temperature at which the fuel begins to solidify. The degree to which this is a problem depends on the feedstock. Vegetable oil based biodiesel has been used successfully down to -15°C but biodiesel made from saturated fats (including animal fats) will solidify at higher temperatures. There are wintering agents that can be used for mineral diesel and for biodiesel to improve the cold-weather performance.

EN590 is a standard for mineral diesel which specifies a density of 0.82-0.86, outside the range for biodiesel. Blends of biodiesel and petrodiesel can easily achieve EN590 but it is virtually impossible to reach it with B100 (pure biodiesel). This can cause some confusion with vehicle manufacturers specifying that biodiesel used in their vehicles must reach EN590.

potential for different scales of production

This section examines five different scales of production, and explores their viability. Remember that the Chancellor reduced the duty on biodiesel by 20p per litre in 2002, dependent on achieving an ester content of at least 96.5%.

scenario 1: I want to make biodiesel for myself

...so that I can run my vehicle on a more environmentally-sound fuel (and hopefully save some money)

This is the simplest option to achieve. The Environment Agency explicitly states that the Integrated Pollution Prevention and Control (IPPC) regulations do not apply to an activity that is in connection with a domestic dwelling. Customs and Excise make it relatively easy to register and pay duty. If you want to pay the lower rate of duty then you'll have to make biodiesel that is not less than 96.5% ester content by weight. You should be able to achieve this using the methods outlined in this book. Some biodieselers manage to pay the lower rate of duty without having any tests done to prove they have met this standard. In other cases HMCE will demand accredited test results.

Making biodiesel is a messy and time-consuming business. You will have to be pretty interested in it, and consider it a hobby, or it could become tedious. You will probably be making 100-litre batches, and your total production for the year will obviously depend on how many batches you make.

Costs: you will need a reactor. You could build one yourself; you could attend a LILI 'self-build biodiesel reactors' course; or there are a few companies who sell small-scale reactors on the Internet. You will also be able to buy a 'Goldenfuels' reactor from LILI soon.

Your waste oil could be free, if you develop a relationship with a local chip shop or other cooking oil user; the amount of potassium hydroxide that you will use per litre is small so that it is only an occasional purchase. The amount of methanol required will be around six pence worth per litre of biodiesel produced. You can recover two to three pence back. Then there is electricity, and your time. If you account for your time in financial terms, it may not be worth it, but as mentioned earlier, if you consider it a hobby, then you can make a very cheap, environmentally-friendly fuel. Add to this the feeling of driving past petrol stations knowing that you don't actually need them any more (priceless) and the security of knowing that you can make your own fuel whether or not there are future price rises, blockades at ports and refineries, or queues round the block at forecourts, and you may well feel that it's a very worthwhile thing to do.

There are quite a few people around the UK doing this.

scenario 2: I want to make biodiesel for sale

...as a small-scale business producing around 1000-10,000 litres per week

This is not going to make you a millionaire but there are many other good reasons to operate local small-scale plant.

As soon as fuel is being produced for sale many rules / regulations apply, which don't apply if you are making it on a domestic scale for yourself. An application for an Environment Agency permit will need to be sought at a minimum cost of £2500, although the yearly permit to operate is £400...as long as you can qualify for 'low-impact' status. High-quality fuel will be required so that it can be sold with the lower rate of duty applied.

Local Authority planning consent will be required, but most of the sort of suitable industrial premises you will find yourself viewing will already have the B1 planning application granted. Health and Safety requirements for operating the plant might incur consultancy fees for such services as making risk assessments. The operation would incur the usual insurance, administration costs etc. that are required by any trading company. A more serious approach to oil collection will be needed, perhaps working with other local businesses.

There are 2 or 3 small companies or co-ops in the UK trying to do this. At the time of writing only Sundance Renewables in Wales is up and running. Nobody can say as yet if this sort of business is financially viable in the long term; in the short term it definitely requires resourcefulness and the ability to take time out from sane, paid employment!

scenario 3: we want to form a fuel club or co-op

...as a not-for-profit enterprise providing members with fuel

This might offer a way around the costs of running a full-scale biodiesel business. It would involve a co-operatively owned plant that is operated by people who require fuel for themselves, and where there are no sales of fuel, only the covering of costs. In this situation you might be able to argue for the 'domestic interpretation' of the IPPC (and other) regulations. The fuel could be made for a 'not-for-profit' group and no financial transaction need take place. You would be still be responsible for paying your duty and have to comply with safety and storage regulations. Registering a co-op is cheap and there's quite a bit of help available. Perhaps you could get funding as a community project.

This sort of arrangement is common in the US and elsewhere. We don't think anybody has tried it in the UK...yet!

scenario 4: I want to form a biodiesel purchasing co-op

...to use bulk buying power to get value-for-money biodiesel

This is a good option for a small group of people who want to use biodiesel in their vehicles. Buying a bulk container (an 'Intermediate

Bulk Container' or 'IBC' of 1000 litres) between a group minimises the delivery cost and the cost of the container. You will get the fuel at the cost of normal diesel (or within 5 pence of it) but will have to provide a suitable space for the tank that complies with oil storage regulations. The diesel could be distributed in small portable containers of maybe 20 litres or decanted into oil drums (c. 205 litres but can vary). Containers of known size preclude the requirement for metering equipment. It's worth shopping around - not all the biodiesel being sold is of excellent quality, whatever the websites say. Some of what is being sold as 'biodiesel' is in fact an emulsion of vegetable oil and solvents and should be treated with caution. There have been problems with local Environment Agency officers treating this biodiesel as 'waste' and applying IPPC restrictions to it though this is probably a localized problem.

We have been quoted £8000 for 10000 litres of biodiesel including VAT and delivery.

We know of a few people who have tried this. It works well, when they can get hold of decent and well-priced biodiesel. The problem tends to be that suppliers come and go, and that attempts to make such schemes grow sometimes fail because of the virtual impossibility of making any money out of it. One interesting approach is that of Hour Car in Hebden Bridge who run a car sharing scheme and fuel their vehicles on biodiesel.

scenario 5: I want to start a commercial biodiesel operation

...to make lots of money and save the environment before breakfast

The cost of building a potentially profitable plant, including all the application fees etc. could be somewhere in the region of £500,000 – £2m, but would be less if you designed and specified the plant yourself, and project-managed the installation. You would probably need to make at least 50 tonnes of high-quality product per week, and sell it for as much as you can to one or two customers to keep distribution costs down.

The future price for waste (and new) cooking oil is a major consideration. Now that waste oil has been banned in the production of cattle feed, the cattle feed producers will need to buy new oil at a much higher price. This could possibly increase the production of new oil and the availability of waste oil, but as new biodiesel producers enter the market, there will come a point when the waste oil runs out. You will have to consider the possibility of requiring new oil for your plant when writing your business plan. The viability of this depends on the price of both mineral diesel and vegetable oil.

The more you can produce the safer you are. If you can sell the fuel at a penny cheaper than mineral diesel the market is vast (although mineral diesel prices can vary dramatically in different areas and on different forecourts). The cost of crude oil will almost certainly start to creep up (and rise sharply if there's a shock - as will your methanol). Alternatively, the duty may be reduced again, which will significantly increase the viability of biodiesel production. As with all businesses, you can't know for sure what will happen, and there are risks involved. The price of glycerol alone could be the difference between success and failure: plants have been cancelled when the price crashed after the first large European manufacturers came on stream.

Your customers are most likely to be haulage companies, local authorities or other bulk users, to keep distribution costs to a minimum. Also, many engine manufacturers are still reluctant to offer warranties for biodiesel (even for a 5% blend), so selling to the general public could be difficult. Suitable partners for the enterprise could be an oil-collecting house and a bulk consumer. The plant could be sited at either the oil house or near the user so that transport of feedstock / product is minimised.

regulations

Customs & Excise

To quote an earlier and more compact version of this book, we: "strongly advise you pay your tax – can be found out – smell of chips"

In order to maintain a trouble-free biodiesel production environment we advise you to pay all the duty that you will owe Customs and Excise (HMCE). The penalties can be very severe. HMCE have the right to impound your vehicle and fine you, they also have the ability to incarcerate you if found guilty in a court of law. We have not heard of this happening related to biodiesel, although when someone gets caught using red marked agricultural diesel in a road vehicle they get their vehicle confiscated and then charged for all the miles on the clock for which they cannot account for having paid the full rate of duty.

Another good reason to comply with the law is so as not to enable bad press to arise from the use of biodiesel. Most people remember the media a couple of years ago jumping on the fact that north Wales police were lying in wait to catch people buying up stocks of vegetable oil for use as an untaxed fuel. Another less publicized fact was that one of the people spreading information about the possible use of vegetable oil emulsions in Diesel engines was monitored by HMCE for a period of time. His email correspondence was scrutinized prior to a visit they made to his home. We think this to be an unusual scenario for the most part but a possibility not to be dismissed off-hand.

At the time of writing the duty rates were as follows:

Fuel	Pence per litre
Ultra-low sulphur diesel (ULSD)	47.10
Heavy oil which is not ULSD (i.e. conventional diesel)	53.27
Marked gas oil and ULSD oil not for road fuel use	4.22
Fuel oil	3.82
Other heavy oils delivered otherwise than for use as road fuel e.g. Marked kerosene for heating Aviation turbine fuel Lubricating oil ...but excluding oils within the gas oil or fuel oil definition	NIL
Biodiesel for use as a road fuel	27.10
Biodiesel used other than as road fuel	3.13
Other diesel substitutes	ULSD rate

The current rates are available from
http://www.hmce.gov.uk/business/othertaxes/roadfuels.htm

Customs and excise make it quite easy to set up an "account" with them to pay your duty.

When you get in touch with them they will send you form EX103a. This allows them to enter your premises at any time. You must make transparent records of production figures (that you probably want for yourself anyway). They want to be able to see any delivery notes that you have if you deliver any of the fuel you make. Every month they will send you a HO930 form to account for the fuel made during the month. You have to pay duty on fuel that is "set aside": i.e. as soon as it is made.

If you want to use or sell your fuel to someone who wants to pay 3.13p tax per litre for 'red diesel' agricultural, generator or boat use then the full tax has to be paid then claimed back. Note that there is not currently a colour marker for biodiesel.

Whether you make one hundred litres a month or ten thousand a day this process remains pretty much the same.

The duty on biodiesel is guaranteed to be 20p below that of mineral diesel until 2007. The recent EU biofuels directive shows us the future though. It states that should the price of ordinary petrofuels become so great that it would make biofuels substantially competitive, the member states should increase the tax of biofuels accordingly.

EN 14214 is the new EU standard for the quality for biodiesel. HMCE's description of biodiesel is derived from this standard. HMCE state:

'Biodiesel' means diesel quality liquid fuel produced from biomass or waste cooking oil:

- the ester content of which is not less than 96.5% by weight; and
- the sulphur content of which does not exceed 0.005% by weight, or is nil.

Despite being derived from the standard, this definition has a quite different meaning, as it omits those parts of the standard that specify what kind of esters "ester content" refers to. Therefore this definition covers partially reacted and unreacted oil, even pure SVO. Expect this to change at some point though!

17.5% VAT is payable on biodiesel if you are selling it as a road fuel, 5% if heating fuel. Neither of these tariffs is applicable if you are not registered for VAT of course.

Environment Agency

There is a big difference in dealings with the Environment Agency between those making biodiesel for their own consumption, and those planning to sell it. If you are making it at home for yourself, you don't need a licence or annual permits. If you want a not-for-much-profit business or a profitable business then you have to apply for an Integrated Pollution Prevention and Control licence (IPPC). The Environment Agency has allowed some biodiesel process plants a low-impact form of this licence. If granted, this will cost two thousand five hundred pounds and annual renewal is four hundred pounds. If you do not qualify then the bill will be much larger, and you will need to spend a lot of money on consultants just getting the licence. Global Commodities of Norfolk, Sundance Renewables of Swansea, and Goldenfuels of Oxford have applied for and been granted the low-impact licence.

In order to legally collect waste oil you need a waste transfer licence costing around £160 for two years. This allows you to collect waste and issue waste transfer tickets that larger caterers may demand. You may only carry waste to an approved waste collection facility with a Waste Management Licence. If your plant has an IPPC licence then this will cover the waste management licence aspect.

One particularly relevant piece of legislation is the guidelines on oil storage. These state that oil must always be stored within a bund (an oil-proof trough which makes sure any spillage will not go into watercourses). In theory the bund should be large enough to accommodate a spillage 110% the size of the largest single vessel. It's a good idea to build a bund even for very small-scale biodiesel operations. Even vegetable oil can have destructive impacts on wildlife, and methanol or biodiesel spills could be much worse.

planning permission

A small commercial operation will need B1 planning permission for industrial space. If a commercial plant is located on a farm, you will need to seek planning permission because the farm won't be covered, unless they have already got planning permission for industrial use.

Health & Safety Executive

Like the Environment Agency, HSE will not take a great interest in the home producer of biodiesel. Of course, you will still want to adhere to best working practices, many of which are mentioned in this book. It is a question of educating yourself, and making sure you know what is required. It is largely common sense (see 'health and safety' section), but some aspects of safety are non-intuitive. For example, the static produced when methanol flows through plastic pipes has been known to create a spark and set fire to the methanol!

If you are a commercial producer there are health and safety regulations you will have to comply with. HSE staff could visit at any time and check to see if you are complying. They also produce guidelines but annoyingly enough many of them are not available for free on the Internet. See resources section for books on the safe use, handling and storage of flammable materials, all of which can usually be found in the reference section of your local library. You will need to employ a trained engineer to check your designs for safety issues. Note that unlike the Environment Agency, HSE have no formal requirement for you to undergo an assessment before building the plant. They do however ask you to register with them using Form F9, Notification of occupation - https://www.hse.gov.uk/forms/notification/f9.pdf

insurance

A small commercial operation will need a combined contents fire and theft insurance for premises, personal liability, and product liability (covering you in case your fuel damages an engine). You will also need employer's liability if you employ anyone. We have been quoted £3000p.a. with £30,000 excess for combined insurance for a small plant.

weights and measures

If a commercial operation is selling biodiesel via a pump in a forecourt-style operation, it will need to be calibrated by the local Trading Standards department. The pump will need to be 'type approved' that is, a design for which the patent has been approved by Weights and Measures. Such pumps start at about £4000 new, which means that for small operations, it is much better to sell biodiesel in containers. Suitable pre-measured containers might include a twenty-litre jerry can, a 205-litre barrel, or a 1000-litre intermediate bulk container (IBC). Whichever way you choose to sell your fuel, you can be prosecuted if it is found that you have given 'short measure'.

the future

The long-awaited future has now become the present. To make a real difference to our planet, merely changing the fuel we put in our engines will not do much but perpetuate the ultimately unsuccessful industrial experiment into which we have been born.

The obvious, but hard-to-swallow fact is that to reduce the depletion of the earth's resources, along with the associated pollution and climate change, we need to change our lifestyles to reduce energy consumption. Within our lifetimes there will only ever be more and more people wanting more and more things, but there are no more planet Earths to maintain this accelerating use of resources.

This book and the courses we run have been designed mainly to enable a few people to empower themselves and others to make practical steps to reduce their footprint on the earth.

We are often told that there is not enough land to grow crops to replace all petrodiesel with biodiesel. It has been estimated that for Russia to replace all its diesel use with biodiesel, it would have to use 7% of its land to grow the crops needed – a huge land area, but feasible. This figure rises to 30% in the USA, and over 100% in the UK.

However this should not stop us making a start with what we have already, especially making sure we tap the potential of WVO as a feedstock. Biofuels will have to be used in combination with fossil fuels, and with other renewables as fossil fuels run out. Until that happens, the main use for biodiesel may be as a lubricity additive in mineral diesel, and as a home-made alternative to the petroleum industry for a relatively small number of enthusiasts.

We are demonstrating a real and totally practical solution to the biggest environmental threat of our times, and can only hope that we inspire others to start taking environmental and ethical problems seriously enough to actually do something about them.

A journey of a thousand miles begins with a single step. Good luck!

resources

biodiesel manufacturers / suppliers

- www.biodieselfillingstations.co.uk - lists most of the places in the UK you can get biodiesel at the pump – either 100% biodiesel or in most cases 5% blends
- Global Commodities (East Anglia) (large quantities only) 01362 821582 e-mail oilbio@aol.com
- Rix & Co. (Humberside) 01482 338967 (can supply 1000 litre)
- BIPSE (West Midlands) 0121 544 2333 (not producing at time of writing)
- e-diesel 01606 301222 e-mail ediesel@ebony-solutions.co.uk (1000l minimum) www.ediesel.co.uk
- LILI will be supplying biodiesel in 1000 litre and 205 litre containers soon.

equipment

biodiesel plant

- LILI supply domestic biodiesel reactors, and run a self-build biodiesel reactors course – see website www.lowimpact.org, or call 01296 714184
- Kelly's Industrial Pages – to find kit to build your own – www.kellysearch.com
- www.eurobiodiesel.com – medium to large-scale equipment suppliers, potentially offer other services
- http://biofuelsystems.com - suppliers of various bits including wintering agent
- pumps, tools, trolleys – Machine Mart (nation-wide) 0115 956 5555, www.machinemart.co.uk
- electrical & mechanical components – RS Components - rswww.co.uk

- plumbing, tools, fixings, safety wear – Screwfix Direct 0500 414141, www.screwfix.com or Toolstation, www.toolstation.com

oil drums, tanks and containers

- JE Jones Ltd - 0121 356 9169
- The Tank Exchange - 0870 7080800, www.pirnet.com
- Thornton International Ltd - 01233 740009, www.thorntoninternational.co.uk. stainless steel tanks to 20,000 litres (second-hand, containerised)

second-hand chemical plant

- Ayton Equipment - 01642 711455
- Central Bottling Doncaster - 01302 711056

turnkey biodiesel plants

- turnkey: supply comprises the engineering, hardware, construction and commissioning of the plant
- EIA Warenhandels GmbH - www.biodieselaustria.com
- Superior Process Technologies - www.SuperiorProcessTech.com
- Westfalia Separator - 01908 313366

lab equipment

- RS Components - rswww.co.uk

waste oil

- A list of collectors / suppliers can be found at http://www.recycle.mcmail.com/fats.htm
- www.defra.gov.uk/environment/waste/management/doc/pdf/waste_man_code_leaflet.pdf - an explanation of the waste management ticketing scheme
- keep it local, and reduce the distance the waste oil has to travel - have a word with your local chip shop or kebab van

winterising agents

- http://biofuelsystems.com/wintron.htm Wintron XC30

- www.powerservice.com/arcticexpress_biodiesel_antigel.asp
  Arctic Express Biodiesel Antigel

- www.lubrizol.com/EnvironmentallyCompatibleFluids/lz7671.asp
  Lubrizol 7671A

chemical suppliers

- Hays - 01582 560055

- Jennychem - 01634 290770

- Monarch - 01795 583333

- Performance Chemicals - 0207 231 3737

- Basic Chemicals Ltd – 01494 450701

recipes

- Mike Pelly's biodiesel method, on www.journeytoforever.org

- 3-stage method by Alex Kac, on www.journeytoforever.org (for commercial plant – uses sulphuric acid to re-bind free fatty acids to the glycerine)

- www.webconx.com/biodiesel

- From the Fryer to the Fuel Tank (see books)

laboratory testing services (accredited)

- ASG Analytik-service (Germany). +49 821 486 2518 - info@asg-analytik.de

- Reading Scientific Services 0118 986 8541 – ester content tests

- BSI Inspectorate, Oil and Petrochemicals Division - 01708 631101, e-mail patrick.periam@inspectorate.co.uk

- Advanced Chemical Laboratory 01224 42112

Customs & Excise

- HM Customs and Excise national advice line: 0845 010 9000 - www.hmce.gov.uk

- www.northwales.org.uk/bio-power/howto.htm - how to register as a substitute fuel producer

Environment Agency

- www.environment-agency.gov.uk - general enquiries 0845 9333111

Global Commodities have successfully applied for a low impact IPPC permit. You can get a copy of the application from the Peterborough office although they may charge an admin fee: Environment Agency Regional Head Office, Kingfisher House, Goldhay Way, Orton Goldhay, Peterborough PE2 5ZR - general enquiries 0645 333111. Low Impact IPPC permits have also been obtained by Sundance Renewables in South Wales and Goldenfuels in Oxford

- Pollution prevention guidelines.

www.environment-agency.gov.uk/business/444251/444731/ppg/

PPG02 relates to storage of oil, PPG26 to drums and IBCs, PPG07 to fuelling stations and PPG03 to oil separators. There is also guidance on concrete/masonry bund construction.

health & safety

- Health & Safety Executive information line 08701 545500 - www.hse.gov.uk

- HSE Books 01787 881165 - www.hsebooks.co.uk

- *The Safe Use and Handling of Flammable Liquids,* HSE Books, £8.50

- www.hse.gov.uk/pubns/indg136.pdf - for an introduction to COSHH (control of substances hazardous to health).There are a number of other free leaflets available for download.

- www.coshh-essentials.org.uk - on this site you enter details of your process - i.e. what chemicals you will use and what you will do with them. The site tries to point you to the relevant resources. To use it you will need to refer to MSDS (materials safety data sheets)

- http://physchem.ox.ac.uk/MSDS/#MSDS - a comprehensive alphabetical archive of MSDS for every conceivable chemical.

To understand the risk phrases use this:
http://ptcl.chem.ox.ac.uk/MSDS/risk_phrases.html

- www.ilo.org/public/english/protection/safework/cis/products/icsc/dtasht/index.htm - International Labour Organisation chemical safety datasheets

planning permission

- www.hmso.gov.uk/si/si1987/Uksi_19870764_en_2.htm - explanation of planning permission categories.

weights and measures

- www.nwml.gov.uk/ - national weights and measures laboratory
- www.nwml.gov.uk/legis/regs/liquid.asp - The Measuring Equipment (Liquid Fuel and Lubricants) Regulations 1995

insurance

- Grant Smith Ltd   78, Old London Rd, Brighton, East Sussex BN1 8XQ Tel: 01273 540410.   These brokers have found quotes on small biodiesel plants for public, employers and product liability
- F.M. Green   39 Market Square Witney, Oxon OX28 6AD Tel:01993 701300

trade associations

- Allied Biodiesel Industries (UK) - www.ukbiodiesel.biz
- British Association of Biofuels & Oils (BABFO) 01406 350848 - www.biodiesel.co.uk
- British Biogen – www.britishbiogen.co.uk – trade association for biofuels in the UK
- National Biodiesel Board (US) – www.biodiesel.org

engine warranties

- www.channel4.com/4car/buying-guide/faq/biofuels/biofuels-7.html   channel 4 website - check if your car is biodiesel compatible, and if warranties will be affected by using biodiesel

engine conversions for straight veg. oil

- LILI is running a veg oil engine conversion course – see www.lowimpact.org for details

- Goat Industries (Wales) - www.goatindustries.fsnet.co.uk, email paddy@goatindustries.fsnet.co.uk - two tank conversion

- Elsbett Gmbh (Germany) - +49 (0)9173 77940, www.elsbett.com - new oil only

- www.vegoilmotoring.com – West Wales, 01239 698237; conversions; information; links includes links to other straight oil sites

- www.dieselveg.com – Wolverhampton, 01902 450001; undertake conversions, and sell veg oil with road fuel tax already paid for 66p per litre (includes tax invoices)

books

- *From the Fryer to the Fuel Tank,* Joshua Tickell, £19.95, available from Eco-logic Books - 0117 942 0165, books@eco-logic.demon.co.uk

- *The Biodiesel Handbook*, CS Garnett, £7.50, Chris.Garnett@Tiscali.co.uk

courses

- LILI – how to make biodiesel - www.lowimpact.org, 01296 714184

- LILI – self-build biodiesel reactors, see above

web

- http://biodiesel.infopop.cc - biodiesel forum: to discuss biodiesel with other enthusiasts from across the globe. A lot of expertise is on hand with forum sections on manufacture, business, and a UK section in Biodiesel Connections. Use the find facility to locate topics of interest (for example, try 'continuous process') - a superb resource

- www.jouneytoforever.org/biodiesel (a US site) - a very informative and comprehensive information source

- www.veggieavenger.com/avengerboard/ very useful site with many photos of small scale reactors.

- www.veggiepower.org.uk

- www.northwales.org.uk/bio-power - network of small producers focusing on emulsions/blends. newsletter to sign up for, and a huge links page

- Alternative Fuels Data Centre – (www.eere.energy.gov/cleancities/afdc/) – a US site containing contacts, and lots of information on all types of alternative fuels

- www.biodieselfillingstations.co.uk – a list of the places around the UK where you can buy biodiesel or biodiesel blends

- www.dancingrabbit.org/biodiesel – a US community into making biodiesel; includes information and recipes

- www.webconx.com/biodiesel – loads of practical information, including recipes

- www.sundancerenewables.org.uk - story of one group's attempts to set up a small co-operative plant in South Wales

- www.greasegypsy.com/biofuel.html – biggest links page we found on biofuels

appendices

appendix 1: technical description of biodiesel

- common name: biodiesel
- chemical name: fatty acid methyl esters (FAME)
- formula: C14-C24 methyl esters
- boiling point: >204°C
- vapour pressure: <5mm Hg @ 22°C
- specific gravity: 0.87 @ 25°C
- specific heat capacity: 2.47KJ/Kg/K (rape methyl ester)
- melting point: -1°C
- evaporation rate: less than 0.005 (butyl acetone = 1)
- solubility in water: insoluble
- appearance and odour: light to dark yellow clear liquid / pleasant odour
- flash point: 160.5°C (PMCC method)
- flammable limits: n/a
- extinguishing media: dry chemical, foam or CO_2 – treat as oil fire
- reactivity: stable
- materials to avoid: strong oxidizing agents
- hazardous decomposition / by-products: carbon dioxide, carbon monoxide
- hazardous polymerization: will not occur

appendix 2: fuel options

There are two major non-biodiesel options for using vegetable oil in a Diesel engine:

emulsions

As we know, vegetable oil is too viscous to use directly in an unconverted Diesel engine – we have to reduce the viscosity. This can be done with a solvent such as white spirit. It's usually done to a ratio of 6-8% solvent by volume of vegetable oil used. Used vegetable oils can be collected and cleaned for this use. There have been some instances of people saying that it is suitable to use methanol as the solvent. While it is true that it reduces viscosity, it can lead to serious problems. We have talked to someone who was advised to use methanol in his emulsion. A sensor in the fuel pump didn't like the methanol, and it delaminated, so the car would not work. He had to pay eighteen hundred pounds to get it serviced and replaced!

Kerosene costs around 20p per litre, and is delivered to homes for heating. You can't use it in a Diesel engine, because it has very bad lubricating properties, and everything that needs lubrication, like the fuel pump, would die. Some people have used kerosene cut with vegetable oil because it is cheap and easy. It has been reported to work, but you don't know what it is doing to your engine. Don't even think about it in a modern engine, and even in an old engine, if you value your vehicle, it's not worth it.

Although many people are very enthusiastic about emulsions, we do not think they are currently the way to go. There are no studies that we know of into the effect of these blends on engine wear, and everything we know suggests that it will not be positive. Similarly the environmental credentials of these fuels are not clear - there is a total lack of data. It would seem that they would give rise to harmful emissions such as acrolein but nobody is sure. More research is needed, as they certainly seem easier and less energy-intensive to make than biodiesel. There are quite a few people using and promoting emulsions in the UK. They seem to be paying the lower rate of duty at the moment though this may change in the future.

straight vegetable oil

The SVO option is well established in other countries and is starting to catch on in the UK. It involves changes to the fuel system which range from simple to high-tech. In general it makes sense to convert the car rather than the fuel if you have a solid, reliable vehicle which you plan to keep.

two-tank method

An extra tank is added for vegetable oil. You start with diesel, then switch to SVO when the engine is hot. The SVO is heated via a heat exchanger system which uses the waste heat of the engine to reduce its viscosity. The heat exchanger can be as simple as re-routing the coolant through a tube around which the fuel line is wrapped. Before stopping, you switch to the diesel tank so that there will be mineral diesel in the fuel line when you come to start your car again. Cost installed c. £800. DIY kit c. £400. Problems include forgetting to switch to diesel again before stopping, which means that there will be SVO in the fuel line when you come to start the engine again. Also, there may not be room in smaller vehicles for a second tank, which would take up storage space. You can also stop and start on biodiesel.

one-tank method

Elsbett, the German SVO specialists, supply a kit, with a few installers in the UK. It's simpler – you don't have to remember to switch to mineral diesel at the end of your journey. Veg oil and mineral diesel can be mixed in the same tank, but Elsbett say the system is not suitable for WVO use.

Elsbett have also designed an engine where SVO is the fuel and the coolant, so it gets continually heated. When starting from cold, there is an electrically-heated jacket on the fuel filter. Glow plugs stay on until the engine has fully warmed up. The injectors are designed for more viscous fuels.

Different parts of the Elsbett engine can be fitted to existing engines, but this generally involves a sizeable investment.

There are other one-tank systems available at lower cost. These are designed to improve the SVO compatibility of engines that are already suitable for SVO to some degree. That generally means IDI engines made before 1995, with Bosch fuel pumps. Older Mercedes Diesels are excellent candidates.

More information on engine suitability (including the Elsbett) for running on veg oil can be found on www.vegburner.co.uk/suitability.htm.

People running on SVO in the UK have on occasion managed to pay the lower rate of duty as their fuel is indeed >96.5% ester. Hopefully HMCE will clarify this point in the near future.

appendix 3: ILO health & safety datasheets

The following sheets cover the health and safety aspects of methanol and potassium hydroxide. They are included in this book with the kind permission of the International Occupational Safety and Health Information Centre, part of the International Labour Organisation.

These and many other sheets are available from:

www.ilo.org/public/english/protection/safework/cis/products/icsc/dta sht/

International Occupational Safety and Health Information Centre (CIS)

METHANOL

ICSC: 0057
April 2000

Methyl alcohol
Carbinol
Wood alcohol

CH_4O / CH_3OH
Molecular mass: 32.0

CAS No: 67-56-1
RTECS No: PC1400000
UN No: 1230
EC No: 603-001-00-X

TYPES OF HAZARD / EXPOSURE	ACUTE HAZARDS / SYMPTOMS	PREVENTION	FIRST AID / FIRE FIGHTING
FIRE	Highly flammable. See Notes.	NO open flames, NO sparks, and NO smoking. NO contact with oxidants.	Powder, alcohol-resistant foam, water in large amounts, carbon dioxide.
EXPLOSION	Vapour/air mixtures are explosive.	Closed system, ventilation, explosion-proof electrical equipment and lighting. Do NOT use compressed air for filling, discharging	In case of fire: keep drums, etc., cool by spraying with water.

or handling. Use non-sparking handtools.

EXPOSURE

AVOID EXPOSURE OF ADOLESCENTS AND CHILDREN!

	Symptoms	Prevention	First aid
Inhalation	Cough. Dizziness. Headache. Nausea. Weakness. Visual disturbance.	Ventilation. Local exhaust or breathing protection.	Fresh air, rest. Refer for medical attention.
Skin	MAY BE ABSORBED! Dry skin. Redness.	Protective gloves. Protective clothing.	Remove contaminated clothes. Rinse skin with plenty of water or shower. Refer for medical attention.
Eyes	Redness. Pain.	Safety goggles or eye protection in combination with breathing protection.	First rinse with plenty of water for several minutes (remove contact lenses if easily possible), then take to a doctor.
Ingestion	Abdominal pain. Shortness of breath. Vomiting. Convulsions. Unconsciousness. (Further see Inhalation).	Do not eat, drink, or smoke during work. Wash hands before eating.	Induce vomiting (ONLY IN CONSCIOUS PERSONS!). Refer for medical attention.

SPILLAGE DISPOSAL

Evacuate danger area! Ventilation. Collect leaking liquid in sealable containers. Wash away remainder with plenty of water. Remove vapour with fine water spray. Chemical protection suit including self-contained breathing apparatus.

PACKAGING & LABELLING

Do not transport with food and feedstuffs.

F Symbol
T Symbol
R: 11-23/24/25-39/23/24/25
S: (1/2)-7-16-36/37-45
UN Hazard Class: 3
UN Subsidiary Risks: 6.1
UN Pack Group: II

EMERGENCY RESPONSE

Transport Emergency Card: TEC (R)-36
NFPA Code: H 1; F 3; R 0

STORAGE

Fireproof. Separated from strong oxidants, food and feedstuffs. Cool.

IMPORTANT DATA

Physical State; Appearance
COLOURLESS LIQUID, WITH CHARAC TERISTIC ODOUR.

mixtures are easily formed.

e and explosion hazard.

TLV: 200 ppm; as TWA (skin) (ACGIH 1999).
TLV: (as STEL): 250 ppm; (skin) (ACGIH 1999).

Routes of exposure
The substance can be absorbed into the body by inhalation and through the skin, and by ingestion.

Inhalation risk
A harmful contamination of the air can b : reached rather quickly on evaporation of this substance at 20°C.

Effects of short-term exposure
The substance irritates the eyes, the skin and the respiratory tract. The substance may cause effects on the central nervous system, resulting in loss of consciousness. Exposure may result in blindness and death. The effects may be delayed. Medical observation is indicated.

Effects of long-term or repeated exposure
Repeated or prolonged contact with skin may cause dermatitis. The substance may have effects on the central nervous system, resulting in persistent or recurring headaches and impaired vision.

PHYSICAL PROPERTIES

Boiling point: 65°C
Melting point: -98°C
Relative density (water = 1): 0.79

ENVIRONMENTAL DATA

Solubility in water: miscible
Vapour pressure, kPa at 20°C: 12.3
Relative vapour density (air = 1): 1.1
Relative density of the vapour/air-mixture at 20°C (air = 1): 1.01
Flash point: 12°C c.c.
Auto-ignition temperature: 464°C
Explosive limits, vol% in air: 5.5-44.
Octanol/water partition coefficient as log Pow: -0.82/-0.66

NOTES

Burns with nonluminous bluish flame.
Depending on the degree of exposure, periodic medical examination is indicated.

IPCS
International
Programme on
Chemical Safety

Prepared in the context of cooperation between the International Programme on Chemical Safety and the European Commission © IPCS 2000

Updated by AS. Approved by EC. Last update: 10.10.2000

For further information please contact the International Occupational Safety and Health Information Centre
at Tel: +41.22.799.6740, Fax: +41.22.799.8516 or E-mail: cis@ilo.org

International Occupational Safety and Health Information Centre (CIS)

POTASSIUM HYDROXIDE

ICSC: 0357
October 2000

Caustic potash
Potassium hydrate
Potassium lye

CAS No: 1310-58-3
RTECS No: TT2100000
UN No: 1813
EC No: 019-002-00-8

KOH
Molecular mass: 56.1

TYPES OF HAZARD /	ACUTE HAZARDS / SYMPTOMS	PREVENTION	FIRST AID / FIRE FIGHTING

EXPOSURE		
FIRE	Not combustible. Contact with moisture or water may generate sufficient heat to ignite combustible materials.	In case of fire in the surroundings: all extinguishing agents allowed.
EXPLOSION		

EXPOSURE	**AVOID ALL CONTACT!**	**IN ALL CASES CONSULT A DOCTOR!**
Inhalation	Corrosive. Burning sensation. Sore throat. Cough. Laboured breathing. Shortness of breath. Symptoms may be delayed (see Notes).	Local exhaust or breathing protection.	Fresh air, rest. Half-upright position. Artificial respiration if indicated. Refer for medical attention.
Skin	Corrosive. Redness. Pain. Blisters. Serious skin burns.	Protective gloves. Protective clothing.	Remove contaminated clothes. Rinse skin with plenty of water or shower. Refer for medical attention.
Eyes	Corrosive. Redness. Pain. Blurred vision. Severe deep burns.	Face shield, or eye protection in combination with breathing protection if powder.	First rinse with plenty of water for several minutes (remove contact lenses if easily possible), then take to a doctor.
Ingestion	Corrosive. Abdominal pain. Burning sensation. Shock or collapse.	Do not eat, drink, or smoke during work.	Rinse mouth. Do NOT induce vomiting. Give plenty of water to drink. Refer for medical attention.

SPILLAGE DISPOSAL	PACKAGING & LABELLING
Sweep spilled substance into suitable containers. Wash away remainder	C Symbol
	Unbreakable packaging; put

with plenty of water. (Extra personal protection: complete protective clothing including self-contained breathing apparatus).

R: 22-35
S: (1/2-)26-36/37/39-45
UN Hazard Class: 8
UN Pack Group: II

breakable packaging into closed unbreakable container. Do not transport with food and feedstuffs.

Transport Emergency Card: TEC (R)-123
NFPA Code: H 3; F 0; R 1

Separated from strong acids, metals, food and feedstuffs. Dry. Well closed. Store in an area having corrosion resistant concrete floor.

Physical State; Appearance
WHITE, DELIQUESCENT SOLID, WITH NO ODOUR.

Chemical dangers
The substance is a strong base, it reacts violently with acid and is corrosive ium, tin and lead forming a e ICSC0001). Reacts with

ammonium salts
to produce ammonia and causing fire hazard. Attacks some forms of plastics, rubber or coatings. Rapidly absorbs carbon dioxide and water from air. Contact with moisture or water will generate heat (see Notes).

Occupational exposure limits
TLV: 2 mg/m³ (ceiling values) (ACGIH 2000. MAK not established.

Routes of exposure
The substance can be absorbed into the body by inhalation of its aerosol and by ingestion.

Inhalation risk
Evaporation at 20°C is negligible; a harmful concentration of airborne

Effects of short-term exposure
Corrosive. The substance is very corrosive to the eyes, the skin and the respiratory tract. Corrosive on ingestion. Inhalation of an aerosol of this substance may cause lung oedema (see Notes).

Effects of long-term or repeated exposure
Repeated or prolonged contact with skin may cause dermatitis.

Boiling point: 1324°C
Melting point: 380°C
Density: 2.04 g/cm³
Solubility in water, g/100 ml at 25°C: 110

This substance may be hazardous to the environment; special attention should be given to water organisms.

NOTES

The applying occupational exposure limit value should not be exceeded during any part of the working exposure.
The symptoms of lung oedema often do not become manifest until a few hours have passed and they are aggravated by physical effort. Rest and medical observation are therefore essential.
NEVER pour water into this substance; when dissolving or diluting always add it slowly to the water.
Other UN number: UN1814 Potassium hydroxide solution, hazard class 8.

IPCS
International
Programme on
Chemical Safety

Prepared in the context of cooperation between the International Programme on Chemical Safety and the European Commission
© **IPCS 2000**

Updated by AS. Approved by EC. Last update: 18.12.2000

For further information please contact the International Occupational Safety and Health Information Centre
at Tel: +41.22.799.6740, Fax: +41.22.799.8516 or E-mail: cis@ilo.org

notes

ted in the United States
6LVS00001B/69-70